The Ever Changing World

Secrets Revealed

Timothy J. Amdahl

ISBN-10:151692908X
ISBN-13:9781516929085

DEDICATION

I dedicate this book to God, because it is only through God we live here on earth. It is He who has given us all and taken back away. It is not for us to question him. I would like to think it is through him I am able to be used as a tool to bring out what some know, and many do not. Our world is forever changing, and we need to take the time to look, observe, and witness our own. Those that say they are and are not.

I want to thank my wife Rosie who took the time to edit my book thank you very much.

CONTENTS

WORDS TO PONDER

Let not the riddle of yesterday confuse us today, for the riddles of yesterday were written for tomorrow. Know that this book brings out some things that have, is, and will be forever changing in our world.

CHAPTER ONE

America's technology is continually improving as we look at yesterday's science fiction, we see it as today's reality. What once seemed ridiculous, or implausible, now is the norm in everyday living.

Who remembers the television series *Star Trek*? I grew up as one of the kids watching *Star Trek*. Who would have ever thought transportation of that kind could ever become real. Today Microsoft has developed what is known as holoportation. Holoportation is a new type of 3D capture technology that allows high quality 3D models of people to be reconstructed, compressed and transmitted anywhere in the world in real time. When combined with mixed reality displays such as Holo Lens, this technology allows users to see, hear, and interact with remote participants in 3D, as if they are actually present in the same physical space. Communicating and interacting with remote users becomes as natural as face to face communication

The ability to be able to beam yourself in two or more locations at the same time seems crazy, but it is very real. Microsoft has been working on this and while they have accomplished a lot, there are still areas to be improved. You have to wear a bulky facial head gear unit, and you have to be in an area that is set up with the special cameras.

The question that comes to my mind is how similar the effects are when you see someone arrive and depart using the holoportation system. It looks almost identical to how they beamed people back and forth on *Star Trek* decades ago.

NASA is even working with Microsoft to create a virtual Mars 2030 tour. NASA is teaming up with Fusion VR. Fusion is a television cable and satellite Hispanic news and satire channel owned by Fusion Media company that is owned by Univision Communications. It relies on its parent companies news division, Noticias Univision.

NASA is teaming up with MIT Space Systems Lab. MIT (Massachusetts Institute of Technology). It was founded in 1995. Their goal is to contribute through cutting edge research, helping the present and future exploration of space. Their main mission is developing the technology and systems analysis with small space craft, precision optical systems and international space station technology research and development. They are working towards a demo version that will be available to groups like Oculus Rift, Google cardboard, Samsung VR gear, Playstation VR, and HTC Vive.

So as we start this book we see how Microsoft who is very widely known and a respected company known for its computer technology, along with NASA see this information as important and vital to our world.

There is also Biometrics, which refer to metrics related to humane characteristics. Biometrics authentication, or realistic authentication, that are used in computer science, as a way of identification and access control. Biometrics are used to identify individuals in groups that may be under surveillance.

Biometric identifiers are measurable characteristics that are distinct and allow each person's own characteristics to identify themselves, separating them from everyone else. The use of fingerprints, palm veins, facial recognition, DNA, hand geometry, iris recognition, retina, and odor scent are key areas that use the physical data to identify a person. There are behavioral characteristics that relate to the behavior patterns of each individual person, from the way they walk, talk, and express themselves. They use an example of Keystroke Dynamics, which uses the manner and rhythm in which an individual types characters on a keypad. Where dwell time is measured and can be measured from almost any keypad.

Keystroke logging, also known as keyboard capturing is often done so as the person being logged is unaware they are being monitored. They can tell whether you made mistakes and had to back up, along with the speed of how fast you typed. Let us continue on with the more traditional identification systems.

The drivers license is probably one of the most common. Each card has its own identification number making it unique. On the drivers license you will see a person's, birthday, home address, when the card was issued, and when it expires, along with other personnel information. Many have passports which also have pictures of the individual and personnel information. When traveling with a passport it is not uncommon to be asked what your purpose for visiting a certain location, or country is.

I guess if we are going to travel let's go to somewhere fun, where dreams come true How about looking at Walt Disney World, where fantasy and reality intermingle. Walt Disney World Resort is located in Bay Lake Florida. It opened October 1, 1971. It is said to be the most visited vacation resort in the world, with over 52 million people visiting annually. The property covers over 27,258 acres.

It was created in addition to Disneyland, which opened in 1955. The complex was developed by Walt Disney in the 1960's. It was known as the Florida Project. The original plan called for the inclusion of an experimental prototype community, Epcot the second of four theme parks was intended to serve as a test bed for the new city living, transforming something out of nothing.

The Government of Florida created the Reedy Creek Improvement District, it basically gave Walt Disney the standard powers of autonomy of an incorporated city. Walt Disney died on December 15, 1966 before construction began.

The idea was then abandoned, leaving the concept for a planned community behind. Today at Walt Disney World, biometrics are taken from the fingers of guests to ensure the ticket being used is by the same person who purchased the ticket.

As we continue to look at biometrics, we see CITI Group has teamed up with a company called Diebold, it is based in North Canton, Ohio. Diebold is behind the iris scanning biometric technology. The company is currently testing technology that will allow customers to withdraw cash from an ATM, simply by scanning a person's iris.

Some of the consumers were apprehensive on having their iris scanned. David Kuchenski, who oversees Diebold business development of innovation, supposedly stated that people were more willing to have their fingerprints taken before the other.

The process requires one to register their credentials and agree to have their iris scanned, which could possibly be done right there in the terminal.

The machine is called Irving after the author, Washington Irving of *Sleepy Hallow*. The idea of the machine was to remove the screen and card reader speeding up the time, making the new process only a ten second withdraw. In CITI's test, customers sign in to a mobile application and choose how much money they wish to withdraw. Then they scan their iris at an ATM to receive their money. This is a good example of convenience and technology working together in our ever changing world.

In the world of biometrics we have the obvious identifiers, DNA, fingerprints, along with iris scan, but a scent identifier is now among the others. We can differentiate the odor between two or more individuals. Dogs have been trained on identifying items held by one person from another for forensic purposes.

According to the Spanish researchers, smelling devices could be used at checkpoints in a less intrusive manner. That by itself makes it a more valuable tool. A university in Spain claims it is 85 percent accurate.

They are working on biometrics that focus on your voice as well as your ear canal. There may be some skeptics that will say these characteristics have lots of weaknesses, such as a person with laryngitis, or a person who sweats, or is older now. Age being a factor not just with your voice changing, but your body odor being different by your diet changing. These are all good questions, but I think the machines they have created, do not smell like you or I would. I believe they smell more like a dog that can separate each ingredient individually.

When we look at biometrics we see it is not just for security, where we use it to identify a bad guy, or monitor a potential threat. We see Walt Disney using it in their business to monitor their customers. There is Mayor Polisena, who is looking at using biometrics to pay their employees. Their whole town is going biometric. I am not sure what town he is from, but he confirmed this on video, as he explains, or justifies his reasoning.

The Mayor pointed out it helps with data collection, giving an example of an employee using a few vacation days and then being able to look at his next time sheet knowing what time he has left to use. This reasoning really didn't make much sense to me as it had nothing to do with biometrics.

When we look at the top ten companies involved in biometrics, we see companies like 3M Cogent, they are a leading biometric identification solutions provider to governments and legal and commercial enterprises.

They provide the highest quality identification systems, products, and services with the leading technology. Their check in product security, along with border management.

Apple Authentic was acquired by Apple in 2012. The company is a global leader in the designing and manufacturing of mobile devices. It's products range from I Phone, I Pad, Mac, I Tunes, software, and service accessories, and I Pod segments.

It offers its own operating system, IOS, and OSX, various application software services including I Work, and I Life, and Apple TV. It was the first company to launch fingerprint recognition technology in smartphones, which is now being followed by other mobile device vendors, such as Samsung, HTC, and Motorola. I have the ability on my smartphone to unlock it using my own fingerprint.

If we continue on we will see other companies that are well known such as Fujitsu, it is one of the largest IT service providers. The majority of its revenue comes from the Apac region in Uganda. The company is known for its palm secure product, a hand geometry biometric solution.

There are other companies such as NEC, BioEnable Technologies, Image Ware Systems, M2SYS, Mobbeel, Precise Biometrics, and SIC Biometrics. As we move on from biometrics at this time know that our world is continually changing as we try to advance ourselves to the next level, whatever that may be.

If you are ready we will take a look at some technology that many may not be aware of, called Tesla, also known as the death ray machine. Tesla's work on particle beam weapons, it can be traced clear back to 1893, with the invention of the button lamp. In 1896 Tesla replicated the work of William Roentgen, with the discoverer of X-Rays.

Tesla at that time was shooting X-Rays from distances of up to forty feet away from the machine, creating photographs of skeletons. Tesla was also involved in experiments, shooting cathode rays at targets.

Approximately around 1918, Tesla had a laser like apparatus that he shot at the moon. Along with his writings and research of Nikola Tesla, it is obvious that the button lamp had all the components necessary to create a laser beam.

The construction of the lamp supposedly placed a physical substance such as a piece of carbon, diamond, or a ruby in the center, while continuing to pound the area with electrical energy. It would then bounce off the button onto the inside of the globe and bounce back onto the button. **(http://www.bibliotecapleyades.net/tesla/esptesla2. htm)**.

As we look at Tesla Laser weapon system, referred to as (Laws), it is a technology demonstrator built by the Naval Sea Systems Command from commercial Solid State Lasers, (SSL). It includes development and upgrades, providing a quick response for the fleet with an affordable SSL weapon prototype. This allows naval ships and sailors a method to easily defeat small threats from boats and aerial attack, without the use of bullets.

The Naval Surface Weapons Center Dahlgren Division and managed and funded by ONR Naval Sea Systems Command, OSD's High Energy Laser Joint Technology office and supported by U.S. Fleet Force Command. USS Dewey (DDG-105), performed exercises on July 30, 2012 in San Diego. There they did operational tests where they took down aerial targets using the MK15 Phalanx close in weapons system.

It is designed to take out small ships, or aircraft. It could be eventually used against airborne threats, such as missiles and drones. The laser can take out targets in a matter of seconds. Once again our technology is being used to protect our country from foreign and domestic enemies. As we move on from Tesla know we still have much more to learn.

There was a demonstration done on the USS PONCE, where they took out a small boat and fired the Tesla laser weapons system, it not only got its target, but it left the simulated person standing by it unharmed, along with the table the items were setting on. The target was seven medal tubes that were blown off the table in an explosive manner. They did a similar demonstration against a small aircraft as well, with successful results, bringing it down in flames.

They have a weapon called the Active Denial System (ADS). It is a none lethal energy weapon developed by the military. It was designed for area denial, perimeter security, and crowd control. It has been called a ray gun, or heat gun, because it works by heating the surface of the targets, like skin on a humane target. This weapon was deployed in 2010, but the US military withdrew this weapon from Afghanistan before it became operational.

In 2010 law enforcement planned on using this device in California in the prison detention center. The weapon is being looked at for other tactical and mobility options. It currently is a vehicle mounted weapons system only. Police and military personnel are looking at portability as a method of use in the future.

This device, or should I say weapon has been tested on many personnel, and has been very successful at denying protestors the opportunity to protest in a manner that gives them the edge. It again pushes away a confrontation rather than clash with one. This device was designed to save lives rather than take lives, by using a method of avoidance over a physical altercation. It pushes personnel away from a distance, harming no one.

Autonomous Weapons System is another area where our technology has brought yesterdays science fiction to today's reality. Most people remember the show Terminator, where a robot with humanistic characteristics came to life and could make decisions on its own. Today we have such robots in the making. The military has spent lots of money on autonomous robots using them to carry heavy equipment in the battlefield, as well as smaller robots for reconnaissance. We have used lethal drones for years, however the difference with them is that even though they may be unmanned, they are still controlled by us, through humane intervention.

Experts warn though that fully autonomous robots could be deployed in just a few years rather than decades. The concern is building robots that can kill, while taking no direction from humans. They mention these robots in the wrong hands could kill innocent people that the robots deem a threat. This makes me think of Jade II. It is an AI software system using quantum computing. Quantum computers are different from digital electronic computers based on transistors.

It has the ability to collect immeasurable amounts of data on humans to generate a humane terrain system. They talk about this in their PDF. In geographical location used to identify, eliminate targets, insurgents, rebels, or any labels that can be flagged as targets on a global information grid.

While we may not have the autonomous in full function now, we do have weapons that are automatic that can determine whether or not to engage on a set of parameters.

Some of these weapons are the Phalanx CIWS and the Aegis Combat System. The Phalanx can operate on its own once the parameters are entered in.

Dale Stephens, who is an Associate Professor at the University of Adelaide, has a video out where he talks about Autonomous weapons. He states that, "While we do not have them today, we will probably have them tomorrow, or in the very near future."

There is the land version of the Phalanx CIWS, known as C-RAM (Counter Rocket, Artillery, and Mortar). It is designed to intercept incoming rockets, mortars, and artillery. He also mentioned one called Iron Dome, it is a mobile all weather air defense system developed by Rafael Advanced Defense Systems and Israel Aircraft Industries.

If we look at some of the definitions of Autonomous weapons, we see a weapon system that once activated can select and engage targets without further assistance from humans. The sophisticated combination of sensors and software, can learn to adapt their functioning in response to changing circumstances.

I know I mentioned it earlier, but the similarities between this and Jade II, where computers will or can fix problems on the fly without human interaction is already here.

The SGR-A1 is a robot built by Samsung, who is known for building phones and laptops. They built the SGR-A1 which is a weapon that can kill you. It is completely controlled by human interactions.

This machine was designed to be placed between the North and South borders of Korea. It operates by human command, but in the future could operate on its own.

There is the Taranis project (Raptor), developed out of the United Kingdom. It is designed to carry out a large part of its operation without human interference. It can go into a combat zone loaded with multiple weapons. It can attack aerial or ground targets. It has a low radar profile. It is said to be one of UK's most advanced aircraft.

Taranis a look at the future

This is the UK version of technology, however each country has something to offer when it comes to technology. As we continue on know that this technology is everywhere, worldwide.

Moving on we look at another program called **HAARP**, (High Frequency Active Auroral Research Program). It was an ionospheric research program jointly funded by the U.S. Air Force, U.S. Navy, University of Alaska, and the Defense Advanced Research Projects Agency, known as (DARPA).

Its purpose was to analyze the ionosphere and investigate the potential for developing ionospheric enhancement technology for radio communications and surveillance. The program HAARP operated in Gakona, Alaska on an Air Force base, or site.

The most prominent instrument at the HAARP station is the Ionospheric Research Instrument, (IRI). It is a high power radio frequency transmitter used to temporarily excite a small area of the ionosphere. Work on the HAARP station began in 1993. The current working IRI was completed in 2007. BAE Systems Advanced Technologies was the prime contractor.

As of 2008 HAARP had incurred close to $250 million in tax funded construction and operating costs. In 2013 the program was temporarily shut down waiting on a change of contractors. In May of 2014 it was announced it would be shutting down permanently. Later in that year ownership of the facility was transferred to the University of Alaska Fairbanks. This information came from Wikipedia.org. As we continue on looking at HAARP closer, we will find some see this program as more than what it is perceived as. Some believe there is a conspiracy here too.

Without the ionosphere we would all be fried, it is like a protective bubble that protects us from the scorching sun. Nikola Tesla was a Serbian American inventor, electrical engineer, mechanical engineer, and physicist. He is best known for his contribution to the modern design of alternating current electricity supply system. He experimented with both high and low frequencies, along with electromagnetic waves. He envisioned alternating the weather and creating a shield to protect us from missiles. Nikola claims he knew how to split the earth in two.

ARCO approached Dr. Bernard Eastlund, Plasma Physicist, in 1984 to find a use for the natural gas on the North slope of Alaska, which they could not sell. They wanted him to file an application for enough gas to produce all the electricity in the United States for a full year. The gas was used to power some huge antennas.

The antenna's beam radio frequency energy up into the atmosphere and create on a small scale what the sun normally does. Their reason for doing this is because when you have disturbances in the atmosphere you can't communicate with the satellites.

Congress in 1990 directed the defense department to explore the Aurora regions for communications and navigation and surveillance. The Navy and Air Force managed the program. In the application discussed in the paten were destroying missiles, communication control, and disruption. Other areas discussed, to possibly modify weather, and finally to lift a portion of the atmosphere further out into space, where hopefully it could deflect incoming missiles.

Brooks Agnew, Earth Tornographer, did radio tornography using thirty watts of power looking for oil in the ground. He found twenty six wells throughout nine States with one hundred percent accuracy. HAARP uses a billion watts beaming straight into the ionosphere. If they beamed it at the earth the vibrations would be so strong as to create an earth quake.

HAARP can choose directions moving from one part of the ionosphere to another. You can change the frequencies. Dr. Bernard Eastlund chose a phased ray antenna because it can be aimed at a specific spot.

HAARP can paint designs in the sky, through the energy it creates. While some say it is a completely safe machine. Brooks Agnew mentions that while it is pushing the ionosphere eighty miles out into space with a high energy beam, it's heating up. All those molecules in that ionosphere region are absorbing energy out of that radio beam. If they use the right frequency to push that plum out into space, that energy may discharge back out of the ionosphere back down to earth. It would be about a hundred times the energy released out of a single thunderbolt.

Dr. Nick Begich, talks about when you lift the ionosphere up for even a short time the lower atmosphere rushes in to fill the void, which changes localized weather patterns. During a U.S. Congress subcommittee hearing it was brought up that the U.S. government has created weather tampering techniques so that the New World Order will be able to starve millions of Americans, and control the rest.

During the subcommittee hearing they mentioned they had weather altering equipment even back during the Vietnam War.

There are some that point out the best way to reduce a population is by starvation, or force a famine in an undetectable manner. On the following web website, **https://www.youtube.com/watch?v=SToVBicIrJU** there is mention of earthquakes it was translated from Persian language.

In a small town called, Bandar Lenge on the Persian Gulf. They experienced continuous earthquakes reaching point 3 on the Richter scale. These went on for close to two weeks. They mentioned the water turning a bright red color. Many fish died and some appeared burnt. After a few days a sewer smell covered the area. The local news reported it was due to a toxic seaweed. The town was warned not to consume the fish.

HAARP is what some believe is responsible for the death of a hundred and fifty two creatures, some of which appear burnt on two separate incidences in the Persian Gulf.

As we look at this technology we have to remember that we are talking about the world and its ever changing dynamics from one end of the spectrum to the other end. Are we witnessing technology that will better help us, or will it be the invention that brings us crushing to our knees? What could cause a fish to get so hot in the gulf it would become burnt?

Technology meets life and death victims

As we continue on keep an open mind. It is important to learn everything you can. This is just one example of the many creatures that were allegedly killed.

On the same video link on a previous page, a Mr. Benjamin Fulford spoke of an ordeal he had. Benjamin Fulford is a Canadian journalist who used to work for Forbes magazine. While he was interviewing the former Finance Minister, Mr. Heizo Takenaka. Benjamin Fulford stated, "I confronted him with evidence that he sold out the Japanese financial system to a group of financial companies controlled by David Rockefeller and the Rothschild's. The very next day he got an E-Mail from someone saying he was introduced to Mr. Takenaka. He stated to Benjamin he wanted him to meet somebody. The person Benjamin stated he met, gave him a Free Mason Badge.

This gentlemen said to Benjamin he was a professional assassin, and that he could either stop exposing people, or continue exposing people and die at the age of 46. He stated this assassin told him he could join the Free Masons. He then stated he probably would have had to do that, but he met an honest genuine high level Free Mason, who told him there are thirteen more levels above the 33rd degree in Free Masonry. He stated these people were God and that there was no God.

He asked what their plans were on killing people. He stated the assassin replied, yes. That there are far too many people on earth. Then he said we need to get rid of a few billion people, war does not work, so we use disease and starvation.

Benjamin Fulford stated he believed there were many good Masons out there that felt they needed to join but were scared, hoping someone could come and overthrow the organization.

The very next day after meeting with the Ninja assassin he was contacted by the Asian Secret Society. He stated that they offered him protection. Benjamin stated that they have over six million members. He stated with their protection he continued investigating David Rockefeller.

Benjamin Fulford received a response back from his earlier ultimatum from the supposed ninja assassin, who stated that there would be an earthquake in Japan. Two days later there were earthquakes on two consecutive days right on top of Japans largest nuclear power plant. The city was Niigata Japan.

Benjamin commented that there was a big ball of plasma that had been videotaped ten minutes before the earthquake. Both earthquakes were 6.8 magnitude.

This is where HAARP comes into play, it has the ability to create earthquakes at desired locations by aiming the HAARP machine at its target and shooting a billion watts into the ionosphere rebounding them back down at its desired target, creating mass casualties and damage. This same machine can heat up subterranean water.

Benjamin compared it to the microwave heating up water. He mentioned if you put a billion watts into a storm you could make it much bigger. They are capable to create cyclones, tsunamis, and earthquakes.

After the earthquake members of a certain family stated that their big boss was George Bush Senior, in other words they work for Skull and Bones. He predicted lights would appear before the earthquake on video, which they did. He had also mentioned that Taiwan east satellite detected a fifty percent drop in the ionosphere, in the electric energy above the earthquake zone.

Benjamin Fulford stated this machine has already killed 500,000 people. He mentioned the tsunamis, in Indonesia may have happened due to the political timed events, such as they asked the Indonesia government to open the straits of Malacca to help with terror. They said no then they were hit with a tsunamis and suddenly they were on board with helping. In Myanmar they were about to have an election and just before the election they have a tsunamis. He mentioned conveniently the U.S., British, and French were standing by with food and supplies.

The supplies were already on hand. Benjamin Fulford stated it should have taken a couple weeks to get there yet they were almost immediately ready.

Another link I wish to share shows unusual cloud patterns that show what appears to be caused by HAARP.
https://www.youtube.com/watch?v=YsCRzSKP-p8

Does the circle of life involve HAARP? Here we see a cloud formation that looks manmade along with evenly proportioned clouds. Here are a few more pictures that will make you wonder if there is something more, or if it is just all in our imagination. If you are ready we will look up in to the wild blue yonder at what can only be seen as spectacular and a little scary.

When all we see is not as it is

As we end this section on HAARP we can only wonder what is real and what is fake. Do we live in a world that none of us are truly aware of? As we become more aware of our surroundings, the more we will be witness to our world forever changing.

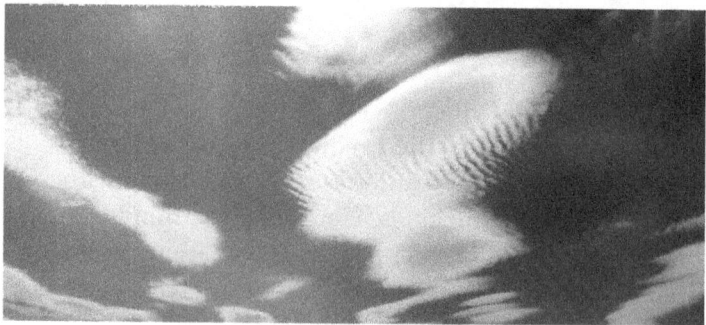

Could these ridges in the clouds be from vibrations from HAARP?

CHAPTER TWO

As we head into the next chapter, we may want to head back in time and remember some of our past. In order to truly gauge something, you need something to compare it to.

The telephone is a good start, for it has been a primary means of communication with families and friends for a long time.

The earliest mechanical telephones were based on sound transmission through pipes or other physical media. The acoustic tin can phone, also known as the lover's phone has been around for centuries.

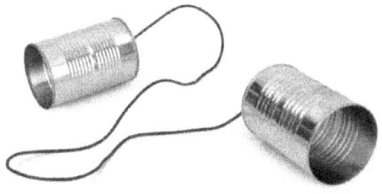

It connects two diaphragms with a taunt string or wire, which transmits sound by mechanical vibrations from one to the other along the wire. Examples of this is two paper cups or tin cans with a string held tightly.

Robert Hooke, British physicist and polymath, did experiments from 1664 to 1685. In 1667 the making of an acoustic string phone was attributed to him.

The first working telegraph was built by the English inventor, Francis Ronalds in 1816 using static electricity.

1915: First U.S. coast to coast long distance telephone call, ceremonially inaugurated by A.G. Bell in New York City and his former assistant Thomas Augustus Watson in San Francisco, California. This was 101 years ago.

If we look at fifty years ago, we see we had the rotary phones. You called someone by rotating the number pad that had holes in it for your fingers. Each hole represented a number. Phones back then were primarily for the house hold, or a business as the phones were connected to landlines. If you wanted to talk to someone on the phone you had to call from your home or someone else's place. You could use a phone booth alongside the road or at a gas station as well.

If we take a few moments and look at these inventors that have contributed much to our world. We will see ordinary men with extraordinary thoughts that pushed the boundaries of everyday thinking. We must never forget where we once were, and who helped get us there.

<u>Antonio Meucci</u>, 1854, constructed telephone-like devices.

<u>Johann Philipp Reis</u>, 1860, constructed prototype *'make-and-break'* telephones, today called **<u>Reis telephone</u>**.

<u>Alexander Graham Bell</u> was awarded the first U.S. patent for the invention of the telephone in 1876.

<u>Elisha Gray</u>, 1876, designed a telephone using a <u>water microphone</u> in Highland Park, Illinois.

<u>Tivadar Puskás</u> invented the <u>telephone switchboard exchange</u> in 1876.

<u>Thomas Edison</u>, invented the <u>carbon microphone</u> which produced a strong telephone signal.

If we look at today's inventors compared to back then it seems they have no names or faces, literally speaking that is. It seems while there are some like Bill Gates whose name everyone seems to know, There are others who seem invisible, hidden behind names like Apple Computers. Which was founded on April 1, 1976 It was founded by Steve Jobs, Steve Wozniak, and Ronald Wayne. Other companies as well seem to focus on their corporate names vs. the inventors name such as Samsung, BlackBerry, NOIKIA, LG, HTC, Sony, and Toshiba.

There are many more examples out there, but the point is how our world has changed from great inventors, to great inventors being bought and owned by big corporations. Places that buy patents and keep the rights, so as to prosper even more.

As we continue on with the phone we can look just at my generation. I was born in 1962. I did not get my first phone till I was eighteen years old. I was working a full time job and paying my own bills. I did have my parents there to assist me at times, which I was very grateful for. I knew to have a phone I would have to be responsible for the monthly charges.

My parents were a little more old fashioned and stood firm with the rotary and push button phones that were connected to landlines.

I was attracted to the cell phone because it was new technology and the convenience that came with having a phone at my beck and call. Being the new generation, I saw this as crucial to my everyday living. It really wasn't but my age and how the media portrayed this item made the cell phone appear as a necessity. To be successful in life, you needed one.

Let's look at the commercials back in the 1980's in regard to the cell phone. Radio Shack had a phone for just $799.00 it was a big bulky electronic box. Back then people would do anything to own one, in fact people even spent $15.00 on a fake phone just for a status symbol. One lady had sold over 45,000 of them. She stated her motto is not what you own, but what people think you own. Motorola had a cell phone that weighed only 30 ounces.

Today Apple I phone SE weighs 113 grams, which is only 3.9 ounces, quite a difference from back then. This just goes to show you in just thirty some years how much technology has come. The phones back then had really only one purpose, and that was to call or contact to someone. The cartoon show, the Jetsons, is an American animated sitcom produced by Hanna-Barbera, originally airing in primetime from September 23,1962 to March 17,1963, then later in syndication with new episodes in 1985 to 1987 as part of the Funtastic World of Hanna-Barbera block.

Who knew that face to face phone calls would ever become real. Today Face Time is a part of our everyday living. Is the media advancing as well, or are they way ahead of their time as they seem to tell us about inventions long before they ever occur?

The phones now are small, light, and do multiple things, from music, photos, calculate, to open and shut garage doors. You can scan bar codes at the department stores for price checks, even use GPS.

GPS was a system that started out in the U.S. Defense Department around the 1970's, later in the 1990's it got going through commercials. It was designed to locate your position with a receiver. There are twenty four satellites in six orbits with four satellites per orbit. Their altitude is approximately 20,000 kilometers. The radius of the earth is about six kilometers. The satellites travel at 14,000 kilometers an hour, that is one complete circle every twelve hours. Usually four to six satellites are visible at any given time. These satellites send time and location information to your receiver and can pin point your location with very close accuracy, between five and ten meters.

Now that you got a basic of the GPS system, let's look at how GPS is being used in our everyday lives. We have cars with GPS built right into them.

All you have to do is punch in an address and hit go. It tells you which way to go, how long to get there, down to the minute. We have GPS tracking on phones letting us and others keep track of your current location, as well as the route you took to get there. We use GPS in tracking our packages we send in the mail every day. Businesses use GPS to track their deliveries as well, making sure their products arrive on time to their scheduled destination.

Mercury Marine Industry uses GPS on their Skyhook model boats. They hit the button and it keeps the boat in that area, keeping it from drifting. It is there to help the not so familiar boater handle his boat better.

John Deere uses it on their John Deere 90 S690 model, it is a combine tractor. Instead of overlapping six to ten feet when the farmer gets tired or it's dusty, or foggy out, it now overlaps six inches helping the farmer better utilize his time and space. It has over 500 horsepower. It will do about fifty five bushels of corn an hour. It can do about twenty five to thirty acres an hour of soy beans. It holds 400 bushels. Back in the seventies it probably only held around 180 bushels. The machine can navigate itself. It uses less fuel because you're not going over the same areas. I think in the farming industry it is seen as a very good tool to have.

Police have used GPS in bait packages. They place packages around at homes dressed as delivery drivers around Christmas time. In those packages are GPS tracking devices. The would be robbers then think it is just a package for their taking. They think it is an easy steal. When they take the package, they are then monitored and tracked then identified and arrested.

Here we have GPS being used by farmers, postal workers, and other businesses, along with boaters, cyclists, and vacationers who want to travel to areas unfamiliar to them. The police are using it along with our military to fight crime and to provide security. If we look at DHS(Department of Homeland Security), we see they reported about GPS on their Homeland Security web page, as they note the following.

Washington the Department of Homeland Security, Science and Technology Directorate (S&T) announced today the successful demonstration of the Enhanced Loran (E-Loran), a precision-timing technology for financial transactions at the New York Stock Exchange (NYSE).

The E-Loran is a low frequency, high power radio navigation signal that is broadcasted by ground based transmission stations, allowing the signal to penetrate through buildings and provide precision timing indoors and throughout urban environments.

"Accurate position, navigation, and timing is necessary for the function and integrity of many infrastructure sectors, such as the electric grid, communication networks, and financial institutions.", said the DHS Under Secretary for Science and Technology, Dr. Reginald Brothers. Ensuring the continuous and uninterrupted availability of critical information ensures our national security." This came directly from the following Web site.

https://www.//dhs.gov/science-and-technology/news/2016/04/20/st-demonstrates-precision-timing-technology-ny-stock-exchange

As we continue on, we will look at E-Loran closer. Dr. Sarah Mahmood, is with the Department of Homeland Security Science and Technology Directorate. She introduced herself as a program manager at DHS. She stated, "For those that may not be familiar with S&T, we are essentially the eyes and the arm of the entire department of DHS. So as you can imagine our mission space is rather broad. Within S&T we have five visionary goals.

One of those goals is called resilient communities. This effort focuses on maintaining resiliency throughout all hazards, including natural disasters. We want our communities to be more resilient.

The community at that level is kind of like a foundation, or the building blocks of our society. So communities which you are all members of rely on info structure.

The jest of this is to enhance our info structure, whether it be electricity, water, transportation, or communication. We really want to enable our info structures to be more resilient, to be able to withstand events and bounce back quicker after events occur. One of the things we are learning is how our info structures rely on GPS."

Dr. Sarah Mahmood stated, "We do not even know all the places we are using GPS. Some items we have available out there are the following listed below.

LandAirSea LAS-1505: It can be used by parents worried about their kid's safety, or employers who wish to monitor their drivers. It is pocket size and does not require an external power source. It receives data from the twenty four satellites we had discussed earlier, and receives data every second, though it doesn't transmit real time data.

Linxup LWVAS1: Offers unlimited text and e-mail alerts, letting you know if your child is speeding, or driving beyond their allowed locations. This item only works in the USA. It requires a monthly subscription.

Anysun TK103b: It uses GPS data and a Google map extension. While it may be cheaper in price, it doesn't sacrifice accuracy. It has SOS alarm and monitoring functions. It has remote arming and disarming.

Spy Spot Investigations GL300: This item is a small battery powered GPS tracker system, that doesn't require being connected to a vehicle power source.

As we seek to unify our past, present, and future together with scientific knowledge, experience through trial and error, and creative imagination, we will then see what we thought was only science fiction to be reality waiting to be exposed.

Let us look at something that DARPA (Defense Advanced Research Projects Agency) is working on called **Iron Curtain**. It is revolutionary technology designed to save lives of those in armored vehicles.

Many of our enemies use the RPG (Rocket Propelled Grenade) as a weapon of choice for many reasons. It is Cheap to make, it is extremely powerful, and can be fired from almost anywhere. They can penetrate a concrete wall, or a foot of steel. It takes an RPG about a tenth of a second to travel fifteen yards to its target.

A force field with a different name

DARPA is working on a system that will be fast enough, and powerful enough to detect, track and destroy in seconds, just inches from the vehicle, rockets that could destroy an LAV (Light Armored Vehicle). On a remote area in the Utah Salt Flats, scientists prepare to test the Iron Curtain on an armored Humvee, using simulated RPG's.

The first thing to pick up the incoming RPG is the radar, which has an antenna mounted on the LAV's top, to detect any incoming rockets. The second sensor has optical sensors, as the RPG travels in to this invisible shield the sensors detect and can disarm this rocket inches from hitting the LAV.

They use a camera that shoots about 30,000 frames per second so they can see exactly where the RPG gets destroyed at. The LAV is setup as an autonomous vehicle, which is designed to go around the track. This prevents the use of a manned vehicle, where a person could get harmed in the testing phase. The camera showed that the system would have worked stopping the RPG in about one tenth of a second from hitting its target. The simulated rounds were a success, but when they used live rounds they had issues with the system working.

Technology stands still for no one.

In June of 2008 they tested the system again with live fire, this time it worked. Using a vehicle with only 3/8 inch armor, they shot five RPGs at it. The system worked successfully. While DARPA won't go through the details of how it works, we can feel comfortable knowing the added protection is a comforting feeling to our soldiers and their families.

I remember watching the television series *Lost in Space*, it was an American science fiction show about a family called the Robinson's. They were pioneers in space, but their ship goes off course and that's where the journey begins. They traveled through the galaxies and explored lots of planets along with aliens. They also had a force field to protect their space ship. This makes me think of DARPA's Iron Curtain. The show aired between 1965 and 1968 with eighty three episodes.

This again is Hollywood, presenting a system way before its time. If we move to more recent times we can look at the movie *Robo Cop*, it mixes man and machine creating another interesting show, but again, in the present we see scientists working on Exoskeletons. This is virtually a robotic skeleton that a human gets strapped inside the frame of the robot.

The movie was released July 17, 1987, almost thirty years ago.

As we look at the purpose of this creation it is simple. Make man stronger, by making the robot carry the load as man travels along effortlessly. Dr John A. Main stated if you strap on 200 pounds of body armor on somebody right now they are going to walk about 300 feet, with the Exoskeleton they can walk all day.

The skeleton is made of aluminum. It weighs about 200 pounds. It feels heavy till they activate it. Once they power it up, all the weight goes away. You feel strong and can move around easily. It will allow a soldier to lift a hundred pounds with no more effort than lifting ten pounds.

Rick Jamison, an operator of the Exoskeleton said the machine is too powerful. If the machine wants to do something you can't fight it. The robot is designed to amplify the operators moves without blocking them. To read his emotions and amplify it, while making him stronger. It is essentially a robot that is controlled by a human. Once strapped into the robot, you become the robot.

The difference is you still control it through human intervention, unlike autonomous robots, where human intervention will someday seize to exist.

Are we headed to a time, where we are no longer needed and seen as now the expendable merchandise?

The military is adapting the Exoskeleton for supply handling. This tells me it's already out there. Remember as we go on looking at our technology, we must not forget our past, for our past and present define our future. We were all once innocent children just wanting to play and live a carefree life.

We played with toys that always seemed better and more advanced than our parents. Today our military uses one such toy, but make no mistake it is real and it is lethal on the battlefield, it is called the WASP. It may appear to be nothing more than a toy plane, looks can be deceiving, as DARPA's solution is the WASP. It has all the same features found on a full size airplane. It has an altimeter, gyro, GPS, and magnetometer. The WASP has already been used with success in covert combat missions where it can detect snipers, or terrorist hideouts. It is like having an extra pair of eyes that can see from areas we can't.

The plane can get close, seeing for us from a safe distance. It gives us a little extra advantage. In combat he who has the advantage often wins. If the WASP gets out of radio communication, it will go to a predetermined site and wait to be retrieved. The WASP can fly up to forty miles an hour, that means it can cover great distances in a short time. It is our aerial scout. This gives our military a heads up, way in advance compared to the traditional patrolling we still do. It is great in urban areas where snipers could perch, hiding out of site till it's too late.

WASP can travel up to three miles away from the remote operator, and sustain flight for forty five minutes. The WASP was designed to break apart upon impact, or landing. The pieces that break apart can be put right back together and flown immediately again. The WASP is just one toy that should be taken seriously.

As we move on, DARPA not only works on projects in the air, but under water as well, as they have a new tool in their arsenal called the Power Swim. This is a designed apparatus that helps our swimmers swim farther and faster with less fatigue.

With swim fins a swimmer burns 400 calories an hour, but with Power Swim, less than 100 calories are burned.

At the Naval Surface Warfare center in Maryland they put it to the test as they competed against a fin swimmer and a barefoot swimmer. Power Swim beat them both easily. They improved the swimmer with Power Swim from ten, to seventy five percent. Power Swim increases their speed twice as fast and increases their range, or distance by four to five times.

Another instrument they created was TUNS(Tactical Underwater Navigation System). It is a navigation system they can hold and view under water. It lets them see their exact location, it combines a compass and depth gauge with sonar.

DARPA is working to eliminate IED casualties, by creating a machine that can detect mines while keeping the soldier in a safe distance. DARPA's solution is the MAV(Micro Air Vehicle). MAV weighs in at around four pounds, it is designed to hover an area while they look the area over. The MAV can fly forwards, backwards, or tilt to get a look at its target from any angle. It can detect man size targets over 800 feet away in the day light and 400 feet at night. It detects using its heat sensitive camera, seeing foreign objects in the dirt. Between MAV and WASP gaining air superiority is a must in protecting us all.

It is time to move from the little toys that are very much needed to the bigger toys that will someday command our skies and protect us with the ever changing weapons and technologies that are out there plotting against us in our enemies hands.

The F-22 Raptor is just that toy developed by Lockheed Martin, with supersonic speed and air superiority. It is an air to air fighter, as well as air to surface fighter. It can drop precision weapons. This jet is designed to be the first to clear enemy lines. It can go in early knocking out all the enemies air defenses, along with ground targets, paving the way for the rest of the U.S. military to come in and continue the assault.

In 1985 they proposed replacing the F-15 fighter jet with the F-22 Raptor. Stealth was a primary factor in the future of fighter jets.

The first stealth aircraft was the F-117 Night Hawk, while it was good in the stealth department, it lacked the fighter capability. It was unable to carry weapons, it also relied on other support. The F-22 Raptor has combined the best of both the F-15 an the F-117 Night Hawk. The F-22 Raptor can see the enemy with radar while still remaining invisible to them.

Stealth is the new commander in the sky

It can cruise at one and a half times the speed of sound, over 1,000 miles per hour. As we continue to progress so too does the enemy. Another aircraft that will be flying alongside the F-22 Raptor is the F-35 Joint Strike Fighter, known as the ground attack bomber of the future.

It was designed for the U.S. Air Force, Navy, Marines, and Brittan's Royal Navy. It needs only a short take off and can land vertically. It carries heavy weapons externally, yet if it carries a smaller load internally it is almost as stealthy as the F-117 Night Hawk.

The F-35 is expected to cost near forty million dollars, about eighty million dollars less than the F-22 Raptor. The F-35 is expected to replace the Air Force's F-16 and the A10, the Marine's, AVHV, and the FA-18 of the Navy.

We even have smart bombs like the JDAM, which have GPS on them so they know the location of the plane that dropped them to their intended target they are to destroy. This will eliminate unwanted casualties from missed targets, as long as they put in the right information.

Our world is forever changing, from the face to face battles with our enemies to now being miles away. There was also SWORDS(Special Weapons Observation Remote Direct Action System) this was deployed in 2007 in Iraq.

This was a top secret mission, deploying a new type of soldier never before seen or used on the battlefield. SWORDS are killer robots that were going to assist the U.S. military. During the first five years in Iraq, sixty percent of the casualties came from first contact with the enemy. SWORDS is another project of DARPA's. The robots would replace the humans by having cameras mounted on them letting the robot go into harm's way, while we could monitor what the robots saw in real time.

They also had weapons mounted on them where the operator could fire at the enemy from a safe distance using a set of joysticks, one for direction and one for firing. During the Iraq war three SWORDS were trucked in and attached to the Army's 3rd infantry division. Each robot was armed with an M-249 light machinegun. They were located in strategic locations outside of Bagdad. The operator took cover a few hundred yards away and operated the joystick. Reportedly an operator waited for the word to open fire, when they did something went wrong.

It was said the robot then swiveled its gun towards friendly fire. The SWORDS program was then withdrawn from the Iraq war. There is no confirmation from the Pentagon, or any other high officials, as to the reason SWORDS was pulled from the battlefield.

Kinetic the manufacture of SWORDS is working on a new robot called MARS, it is similar to SWORDS as they give a demonstration to those from the History channel. MARS is more heavily armed with an M-40 ml grenade launcher. This machine is still controlled by human intervention.

In a second world war, the Germans had their own robot called the Goliath, it crawled on tracks and was loaded with explosives. It was operated by remote control. This robot was seen as disposable, with the intention of delivering 137 pounds of explosives to it's target and then detonating.

They had over 7,500 of them roll off the assembly line by the end of the second world war. Sixty years latter another robot would arrive, this one had wings and could fly it was called the Predator-UAV.

Let's look at this aerial robot a little closer. The MQ-1B Predator is no longer being purchased by the Air Force. The MQ-1C Gray Eagle is around 6.66 million dollars. The MQ-9 Reaper costs 14.75 million dollars.

The Predator was another machine that was considered disposable. It was used primarily for reconnaissance. It was used in the Afghanistan war to look for Osama Bin Laden. In 2000 the Predator was fitted with the hell fire missiles, where it could attack the enemy without ever being seen.

The CIA had concerns with the leader Osama Bin Laden, using what they called Afghan eyes. They implemented the Predator-UAV to recon the skies and then assault from the sky taking him out. This was their intended goal.

Change is inevitable, for all must get on, or be left behind. This isn't something new, but a part of who we are as we look at our past and present, so as to determine our future.

We need to remember that while we focus on our future, we are living in the present. Everything must harmonize in a manner that benefits us rather than harm us. In the world of science a name stands out that many may not know much about, it is CERN(Counseil Europeen pour la Recherche Nucleaire) translated from French means, European Council for Nuclear Research.

Sometimes we fear that which we do not know, when we should fear that, which we do know.

If you are ready we will take a look a CERN. It is a European research organization that operates the largest particle physics laboratory in the world, established in 1954. CERN is based in the Northwest suburb of Geneva. In 2013 there were over 2,500 staff members and over 12,000 fellow associates, apprentices, along with visiting scientists, and engineers. They represented 600 plus universities and research facilities. CERN is the birthplace of the world wide web as well.

CERN's main function is to provide the particle accelerators and other infrastructure needed for high energy physics research. The particle accelerator is a machine that uses electromagnetic fields to push charged particles to a near light speed and contain them in well defined beams. A particle is defined as a minute portion of matter, such as a tiny bit, tiny piece, a speck, or spot of a physical substance.

An example used to describe what CERN was trying to do, was they were trying to find out what holds matter together. To me this sounds boring and nothing I am even interested in exploring, until I investigated further and found out it will affect us all both physically and spiritually.

Could these be demonic?

A picture is worth a thousand words.

A look inside CERN.

Understand I am no more educated in science than the next guy, but if science is given financial access through not just our government, but other governments of other countries, should we not at least take a look or be concerned?

If you are ready, I will do my best to present this subject and how it concerns us all. CERN is a device that will allow us to examine particles in their initial state. They use glue as an example, showing how they break glue down when it is still in liquid form, before it hardens. What has come out of CERN is what they call antimatter. It was first produced in 1955. Antimatter is exactly the opposite of matter. Hot vs. cold, high vs. low, fast vs. slow, and good vs. evil. Matter is everything you can see and feel, while antimatter is everything you cannot. Antimatter cannot be controlled, only contained.

One gram of antimatter equals about forty five kilo tons of TNT, which is about four Hiroshima bombs, it is very unstable. They describe antimatter as the other dimension, a place that is uncontrollable, hostile, and evil.

The antimatter attracts paranormal activity. This is not me saying this, but those in the field of science. They mention there is a physical affect in the spiritual world in antimatter. Often demonic entities and paranormal activities are attracted to antimatter. In the world of science, do we go blind to that, which we see as spiritual? Are the scientist so focused on explaining away miracles and signs, or have they themselves fallen victim to Satan?

In the documentary they mention balance, which only makes sense. With everything there is an opposite, there is light and dark, good and bad, happy and sad, hope and hopelessness, and of course matter and antimatter.

They talk about antimatter being pulled out from nowhere in another dimension which is nowhere, but everywhere. Once they identify these properties they will be able to pull out as much antimatter as they want. CERN has found out that antimatter is tied to every single life form on the planet, every single life form. They mention a person has both light and dark in them, referring to matter and antimatter. In other words we have both good energy and bad energy inside us.

In the bible they speak of Psalm 23:4, Yea though I walk through the shadow of death, I will fear no evil, for you are with me. Your rod and your staff, they comfort me. I know you may be worried this is turning religious, but it is what it is. Dark matter is connected to the dark matter. Everything is connected and can never be separated. Does this sound like some crazy science? If they bring dark matter into this realm it is still connected to the dark matter from the other realm.

They mention they had to change locations because the antimatter created chaos and Paranormal activity at one of the universities where the antimatter was being contained. They had to move its location because what was happening to the people in the college. People began to have vivid dreams and nightmares. Violent and vile things started happening to people in those areas.

They bring up that dark matter can be brought into this world through people, and though it may be small it is still measurable. It is brought in through emotions, they even know how much a person can handle of antimatter before it possesses them. Not everyone can be possessed.

If containment of a teaspoon of antimatter is not contained it can cause antimatter in another location to activate. We are talking losing containment in New York will cause antimatter in Los Angeles to activate. They mention CERN has a weapon that can harness this and create chaos anywhere in the world, and that they have already used this. Is it possible that the riots displayed by black lives matter protestors in Ferguson, Missouri, in regards to Michael Brown, or Trayvon Martin or the mass amounts of police being shot could all be linked back to this machine?

I see myself as a Christian, as does the person who narrated on one of the videos I watched. He described this as demons who travel through a portal and create chaos. It makes me think about our world in whole, as we have started pulling the word of God out of our State and Federal buildings. President Obama has been brutal to the Christian community, while pandering with the Muslim extremists. The politicians have pushed for transgender bathrooms and gun control laws that favor the criminals, or again the evil ones.

I am adding a couple links so you can check out the videos yourself, just in case you think I misunderstood the information. I will say I found this to be more interesting than I had originally thought, or expected.

A machine from CERN

CERN's large Hadron Collider could be linked to a multitude of earthquakes. Recently in just a two month span there were thirteen earthquakes reaching 6.0 on the Richter scale, or higher. One earthquake happened in Nepal, it reached between 6.6 and 7.8 killing thousands. There was one in Papa New Guinea that reached 7.1 to 7.5. CERN had fired up its huge machine just before the unusual stretch of earthquakes. It was then shut down just fifteen minutes after the last burst of energy was shot through the collider. They state it was shut down due to a weasel getting inside.

A machine that collides particles together at near light speed. Knowing all we know now, makes you wonder as an earthquake could be seen as an explosion, or a form of chaos. If you want to know more, check out Conscious Life News video on youtube.

I believe I have only scratched the surface when it comes to CERN. As we move on understand our world is not the world we thought it was ten, fifty or a hundred years ago. Maybe I am just becoming more aware of the things that were always obvious to others, I am not sure. If you are ready we will continue on.

I think from here we should look straight ahead, or should I say up, or maybe we need to look behind us? It all depends on where you stand. As we continue on we will look at portals and see what science has to say about them.

Explaining the unexplained
PORTALS

Are we to ignore that which is in front of our eyes, or search deeper for a more logical answer?

CHAPTER THREE

As we open one more chapter, we go further into the unknown where portals and worm holes that once seemed as only make believe, are exposed as real and all around.

A scientific researcher from NASA has found hidden portals on earth's magnetic field, which open and close dozens of times daily. These portals create a path that go uninterrupted to the sun's atmosphere over ninety three million miles away.

NASA has labeled these portals as X-Points, or electron diffusion regions. They are located tens of thousands of kilometers from earth. The portals are created through a process of magnetic reconnection.

They use the word boom tube, which is slang for a fictional extra dimensional point to point travel portal, a form of teleportation, like what we talked about earlier.

The difference is though that earlier we were just talking about an image or an illusion, where you appeared to be somewhere else through the work of cameras and technology, here we are talking about the real thing.

A portal was seen in the sky over Norway, as they show pictures of this portal you can see rings, or a spiral effect. They state the photos are real and not photo shopped in any manner.

What do you think?

When reality is introduced to science fiction we can only wonder what is to come, and even then will we believe what's right there in front of us?

Let us get back to these portals before the next one disappears. If we look at what a portal is, we might see our world very differently.

A portal s an opening in space, or time that connects travelers to a distant realm, to a kingdom of another time and place. It is seen as a shortcut to the other side, to a place we thought we could never venture. Jack Scudder is a plasma physicist, and a professor from the University of Iowa. He is the one who discovered the X-Points.

There are places where the magnetic field of the earth connect to the magnetic field of the sun. This creates an uninterrupted path directly from one to the other. Observation by both NASA's spacecraft and Europe's cluster probes show that these portals open and close dozens of times daily. They are generally located tens of thousands of kilometers from earth.

They talk about the geomagnetic field meeting the rushing solar wind. Many of these portals open for very short times, others are immense and sustaining. Energetic particles can flow through the openings, heating the earth's upper atmosphere, creating geomagnetic storms that ignite bright polar aurora, also known as the Northern Lights.

NASA is planning a mission called MMS(Magnetospheric Multiscale Mission). It is expected to be operational in 2014 to study the phenomenon. It will use the MMS to travel around the portal using sensors to observe how the portals work.

The problem is finding these portals so they can surround them. They open and close without warning, until now. Jack Scudder has found the key.

Portals form by the process of magnetic reconnection, mingling lines of magnetic force from the sun and earth crisscross and join to create the openings.

Jack Scudder looked at data from a space probe that orbited earth more than ten years ago. NASA's polar spacecraft spent years in earth's magnetosphere, where it encountered many X-points throughout its mission.

What is important is that we are aware of portals and of time travel. How much do we truly know, I'm not sure. Have we visited other places, again I do not know. These questions can be answered if you ask NASA, the problem is that most the questions you ask, will probably never truly be answered till well after the fact. I will explain shortly.

Throughout this book you will hear me mention God, because our ever changing world is not ours, but Gods. He has just allowed us to live here on his planet in his world. I can give one example that I witnessed myself that science can't explain, or can they?

I was driving home one day from work headed past our high school on Highland street, when suddenly I felt a jolt from behind. I had been rear ended by another vehicle. When I went to check out the damage I saw the other vehicle had a bent license plate and a bolt was ripped off of their vehicle as well. The grill was smashed also. While it is true it was just a fender bender in most eyes, in my eyes it made me wonder. When I looked at my vehicle I didn't see any damage, other than a little plastic piece popped out of place and it snapped right back in to place.

How can science explain two forces colliding and nothing happening to one of them? You would think first you would look at mass, angles, strength, and speed. What if it is not about mass, or angles, but rather angels? What if those who believed in God believed it was an angel that looked out after them. What if the angel got between both vehicles protecting both drivers. He protected the one driver from a whiplash type injury, while protecting the other driver from having to pay financially? God does work in mysterious ways.

Let's continue on as we explore more of our world. Before we go further let's think about how we perceive our world in the first place. A fetus sees its world as no further than a woman's belly, that is its world. Once the baby is born he sees his mother as his world, as other members are slowly introduced his world expands. He sees only what is in front of him, gradually becoming aware of his own surroundings. As a little boy he now has added friends and the neighborhood, along with the community to his world. As the boy grows into a young man his world continues to grow adding States, countries and culture to his world.

The same too is his brain, from only hearing and seeing to recognizing and understanding. Our thoughts are molded and formed by how we are taught, some brains have been altered by chemical reactions due to drugs, diseases, illnesses, accidents, and genetics. If we look at our world compared to fifty years ago, have we changed our thinking, our priorities, have we distanced ourselves further from the ones we love? Could it be nothing has changed and we are just becoming more aware of things that we once took for granted?

I remember growing up an playing with the neighbor kids, life seemed innocent and harmless. We would go out and play all day, only come home at lunch and then right back outside to play. Our parents told us to avoid strangers. Our parents knew all the neighborhood kids and watched out for us all.

Today our children have to check in more frequently and we do not know all the kids that run with our kids. Our parents used to be closer staying friends and informing each other when needed. Today sexual predators are out there lurking in the shadows, even in bright daylight they attempt bold moves to snatch innocent kids. Is this something that has always been, or has it gotten worse?

Schools seem to constantly continue to get more challenging as it should, but it seems kids are grades ahead of where we were when we went to school. If we look at values and work ethics, we will see that back then we as children grew up working. We cleaned our rooms, washed dishes, mowed the grass, and other odd jobs. Today, kids have to be pushed to load the dish washer that washes the dishes for them.

Kids now days seem to be more open about sex at an earlier age. In fact it is all about sex, rather than relationships and dating. Could this be due to the breakdown in the family sector, or is it through a lack of education that we have some where lapsed?

Remember time stands still for no one, but rather allows us to be a part of it.

In today's world we look at everything through a sensory. What I mean here is anything you say can and will be used against you in offending someone. We can't even use the phrase, hits like a girl without someone feeling hurt, or wounded. We have the Indians who have been offended, because we name our ball teams after names like Kansas City Chiefs, or Atlanta Braves. They ignore the thought that people wear these teams names with pride and loyalty. They do not wear them to insult or make fun of. Have we become a society of sensitivity?

We have the gays, lesbians, Bi, transgender people, who all want to stand and be recognized. They want to impose their rights and beliefs on others while not caring about those who are Christians and disagree with the LGBT beliefs.

There are those in the Black Lives Matter clan, cult, gang, or organization, which ever title you want to attach to them. These people too wish to punish people for being everything but black. They want everything to be handed to them and think that working for something is beneath them. They wish to have black heritage month, but refuse to honor those who want to honor their history involving the Confederate flag.

It is interesting how the black lives matter clan who gather to rally for peace and injustice, are the ones who are the most violent, often provoking violence and destruction to their own people and communities. We have politicians who encourage and promote this negative behavior just to get votes at a later time. I call this manipulating the ignorant.

The word ignorant by itself sounds offensive. The truth is ignorance is just lacking knowledge, or awareness in general. It does not mean you are stupid. In our society many can't differentiate between the two. To try and explain this, only identifies you as part of the problem, at least in their eyes.

During the slave years the politicians worked it so as to count slaves in their population to help with election statistics. The point maybe that not just blacks, but those that were born in poverty may not truly have the same start in life that others do. If we think about how we figured out how to go to the moon, build a powerful military, become doctors, lawyers, and scientists, then why can't we figure out how to help our neighbors in the poor communities build safe neighborhoods? Why can't we get crime out of those same areas?

Is it possible that those in the poverty stricken areas need to start and police their own. Our world is changing and it is up to everyone to do their part. How do we differentiate between those in the communities that want a better life and those who don't?

Why don't the Black Lives Matter Group start with their own communities and build them outward and incorporate them into the rest of the communities in a positive manner? Would a poverty stricken community that was safe from crime not be seen as encouraging to those with low income? This might bring in more businesses and create competition and jobs, once again building a stronger community. This could help erase the walls that divide us from racial ethnicity.

Welcome to the Mandela Effect. I know you might be wondering what is the Mandela Effect? It is where things that once were, are no more. Where subtle changes have occurred and we are in the midst of it all. To give you some examples, we have Dorothy from The Wizard of Oz, who I and others remember stated, *Toto I don't think we are in Kansas anymore." Now we see many years later she states, "Toto I have a feeling we are not in Kansas anymore.

There was David Letterman who mentioned the movie Interview with a Vampire and now it states Interview with the Vampire. There were others besides David Letterman that had said and confirmed this. I remember Tostinos pizzas and now it is called Totinos. It is missing the letter S. I understand these may just be little changes, but they were once real and now they have been erased from our ever changing world without a trace. Why did they get erased? To know that once was is now no more is very scary, especially when it means all that we remember may not be as it truly is, or was.

This again all points back to CERN and its scientific research on other dimensions. As we continue on with the Mandela Effect, we need to think back and stay focused on other things that once were. There is Kit Kat candy that used to have a dash between the two words KIT-KAT and now it doesn't. I know you are thinking, so they changed their label. The problem is there would still be signs of the old ones. If you look on the internet you will see they seemed to have vanished as if never to exist. Reddi Whip cool whip now spells whip without the H, Reddi Wip. There are more as we continue.

Our history is being erased right before our eyes

Used to be Tostinos

Used to be Fruit Loops

Used to be Kit-Kat

Used to be Oscar Meyer

Now all that was is now no more. Why have these things changed and why can't we find the old ones? If we look at a few more, we will see more questions than answers.

These are just a few of the
Many brands and logos that have
Changed. Reddi Wip used to have
an H in it

Scott towels used to have an S at
the end of Scott. All these
Items have to make you think
What's really going on?

Are we all just imagining, or is there something to
this? I remember Oscar Meyer because my ex-wife used
to be a Meyer and we joked about her name, either way it
too has now changed. I got some of this information off
of WWW.JacobIsrael.com . He has a book out that talks
more about this. There is also Oreo cookies that used to
have Double Stuff on the label and now only has one F
instead of Two

If we remember these changes, how is it that the owners of these companies don't? Is it possible they do but are afraid to say anything? Have the owners memories too been erased? Why do so many people remember the same things that are no more?

CERN talks about exploring other dimensions, and that there are others parallel to ours. They talk about portals and dark matter seeping into these portals, if even just a little. We already know that they said just a little could be compared to a bomb the size of Hiroshima. We know dark matter is somehow described as demonic and evil.

As I point these things out, think back through your life and ask yourself if you ever remember saying, if I only changed one thing. My wife and I have joked about that many times, but I always knew to change one thing, would change everything thereafter. It is possible all would remain the same, but nothing is for sure.

If you think about if you had married your first girlfriend everything from there on would be different. This all creates a ripple effect. If you throw a big rock in the water you will see a big splash then ripples going outwards. If you throw a small pebble you will see a smaller splash and ripples that are there but not as noticeable. Most the time when we talk about the ripple effect, we talk about what we see, the ripples on top of the water, but what about when the pebble sinks in front of a fish scaring it in another direction then it scares another fish or gets directed towards a bigger fish that may eat it.

All these things happened outside our vision, leaving us totally blinded by what has truly occurred verses what we thought may have happened. As we look at some other changes, know there are still more we have yet to discover.

In the movie *Forest Gump*, many people remember the saying "Life is like a box of chocolates", but now it is recorded in the archives as, "Life was like a box of chocolates". If we look at the movie *Star Wars*, we will see Darth Vader say to Luke Skywalker, "No, I am your father". The problem is, others along with myself remember Darth Vader originally saying, "Luke, I am your father. James Earl Jones, who played Darth Vader even stated that it was Luke, I am your father, so how is it these words have vanished, been replaced, and deleted from our world as we know it?

There was C3PO who many remember as being completely gold, but now miraculously he has a silver leg. Maybe the movie industry diversified in case gold went down and silver went up. I am just throwing in a little sarcasm. Most of us remember Smokey the Bear, however now he is just referred to as Smokey Bear. How about *Snow White and the Seven dwarfs* and the famous saying, "mirror mirror on the wall". Now for some reason as if like magic, it states "magic mirror on the wall". In the movie, *Field of Dreams*, most remember that saying as well, "If you build it they will come", yet it too has been changed to "If you build it He will come". How is it these are not as we remember them as? I know you might be thinking does this really matter, after all we are talking the changing of simple words from cartoons or other movies. The labels on cereal and other brand labels now gone.

We have to at least look at it and ask is there a chance there is more to this? If James Earl Jones own words have been erased would you not be a little concerned? Would not the actor know his own lines? There was the assassination of John F. Kennedy where they show four people in the car with him. On another video clip they show six people in the clip. Which one is right? Would this not be crucial evidence if we are missing two people in a criminal case? It is possible maybe this was altered. I can't seem to find this clip so I can't comment anymore on it other than seeing it once before with my own eyes.

What if an island had disappeared as well, no longer on the map. It just vanished out of thin air. An island near Australia in the 1990's movie *Dazed and confused* was missing on a world globe that happened to get filmed on the set in the movie. Did the island sink, even then it would still exist, even if it were under water. There was also an island called Sandy Island, it had vanished as well. They mentioned about Mongolia being much larger in size then they had remembered it. I wouldn't know on that myself because it was never something I thought about. Having said that how many things have gone unnoticed because of being busy with other things?

In 1954 in Tokyo, Japan a Caucasian man showed up stating he was from Taured. Taured does not exist in our world, in our dimension anywhere. The man had all the proper paper work. He was supposedly taken into police custody, where he then vanished. If none of this worries you because it seems so trivial, then let me add a little more as we talk about words from the oldest book in the world, the Bible.

In Isaiah 11:6 "The lion will live with the lamb", that saying is gone now, as it now says "The wolf will dwell with the lamb". In the bible it states in Daniel Chapter Seven verse 25, "And he shall speak great words against the most high, and think to change times and laws: and they shall be given in to his hand until a time and times and dividing of time." Could they be talking about CERN and its time machine capability? "Judge not that lest you be judged." Mathew Chapter Seven verse 1, now it states "Judge not that you not be judged."

The Lord's Prayer too has changed according to some, stating it now says in the earth instead of on the earth. Mathew Chapter Six verses nine through thirteen. I have checked this out and that is what it says, in the earth rather than on the earth.

It is as all that has been written, is being rewritten. As we continue on, know that Gods words are here for us all and his word will never change.

When the word of God is manipulated and censored by someone far less great than He, we need to worry. God's word is the only word and when words and phrases disappear out of the bible we should be concerned, not to be confused with CERN.

Is CERN the devils instrument? We have looked at words that have not only changed, but have made the ones from our past vanish and cease to exist. Our religion is constantly being challenged by misperception and temptation. As we just mentioned words are being changed in the bible. Trust in God and pray to him that he may hear you.

I am no preacher, so speak to God in your own way. I am just here bringing you information on how our world is forever changing. How is it that the words in my bible and yours are changing? This is something we will all struggle to comprehend.

Let's get ready and look at D-WAVE computers, D-WAVE computers focus on temperatures beyond cold, they focus on matter, not matter like gases, solids, or liquids, but strange matter, like super fluids.

A super fluid is a state of matter found in temperatures below two Kelvin. They show these fluids being stirred and once they get going they will continue forever. They state this is quantum mechanics in action.

Quantum mechanics is a kind of physics where the usual rules don't apply. In quantum mechanics each atom moves in a manner that connects to the next, until they become one mass. The atom could be in this spot, or that spot, or multiple spots at the same time. Vancouver, Canada holds one of the most powerful computers known, called the D-WAVE computer.

Dr. Geordie Rose is the Founder and CTO of D-WAVE. He mentioned there are two of these computers in operation now. One which is located at the University of Southern California. He mentioned these computers have the capability to access other dimensions, which currently our computers can't do. He talked about parallel universe. The other computer is located at NASA with Google working right alongside them. These two teaming up together should tell us something of the importance of all this.

Dr. Geordie Rose spoke about NASA wanting to build computers that are human computers. He believes they can do this. He mentioned the D-WAVE computer makes a sound like a heartbeat pounding like a humans. Dr. Geordie Rose described the feeling as though he were at an alter of an alien God.

The Doctor talked about a bit in a regular computer is either a one or a zero, but in quantum computers it can be either, or both at the same time. This system is backed by the CIA and operates at 459 degrees below zero. On the web page they specifically describe it as a quantum computer that taps directly into the fundamental fabric of reality.

The reality is that NASA, Lockheed Martin, Google are all working with quantum computers. Lockheed Martin is an American global aerospace, defense, security, and advanced technologies company with worldwide interests. It was formed by the merger of Lockheed Corporation with Martin Marietta in March of 1995 according to Wikipedia

Will our knowledge come at a cost similar to that of Adam and Eve?

> *For he who has faith, need not knowledge*
> *For God will be there for us then, as he is now*
> *Just like our shadows never leave us in the shade*
> *We are forever together as God is with us*

We need to look around and ask challenging questions.

Timothy J. Amdahl

CHAPTER FOUR

As we continue on, we will look at another area of technology called cloning. Believe it or not this too, we have accomplished. There is a five month old pug named Momotan she is just one of dozens that have been cloned Her owner is of Asian origin. His original pug was named Momoko. When Momoko died the owner wanted to bring her back to life. So he contacted a company that clones, called Sooam Biotech. David Kim is a Sooam researcher.

David Kim stated they had cloned over 600 dogs and that there are no limitations, such as type of breed, or age, or size of the dog. Sooam is located in South Korea.

The first thing you need to have, to get your dog cloned is money, it will cost close to $100,000.00. This could be a good reason we do not see a lot of people cloning their dogs. The dog you want to clone needs to be alive or dead for fewer than five days.
Once your dog has passed away wrap the dog in wet bath towels.

The next step is to place the dog in the fridge and not a freezer to keep it cool. Hopefully no one is wanting to do a Great Dane, or Saint Bernard, that may require a walk in fridge. Sooam needs live skin cells to clone your dog. This process prevents the skin cells from drying out, or freezing. You then need to take your dog to the vet and get a biopsy sample. They will need to cut out an eight millimeter sample of flesh from the abdominal area of the dog if the dog is a live one sample will suffice, if dead you will want multiple samples to increase the odds of finding live skin cells. After that you need to pack the samples in a Styrofoam container with ice to keep it cool. The package then needs to get expedited to Sooam. It may take a few days to get this package through customs and to the company.

Once it gets to Sooam, they will sterilize the sample and cut it in to pieces. It then gets treated with a reagent and chemically disassociated, separating the cells from the tissue. The samples are then placed into a centrifuge, which allow the scientist to collect the cells and grow them into a growth medium. A few weeks later the company will have the cells needed for the cloning process. Sooam then goes to an animal clinic where they use an egg donor They sedate the dog and then slice her open pulling out her ovaries and collect her eggs. Once they have the eggs they then extract the nucleus emptying the eggs DNA.. They then replace the new DNA from the dead dog into the egg. Normally sperm would be needed to complete this, but using an electro cell manipulator, which activate and fuse together the membrane of the cell causes, or creates fertilized embryos.

After just one minute Sooam has a whole batch of dog embryos to work with. They then grab the second dog and go through a similar process while adding up to fifteen cloned embryos to her. Approximately thirty days later the results should be visible. There is a forty percent success rate. Once they confirm the dog is pregnant it takes another thirty days to give birth.

Sooam can house up to fifty dogs at one time. The dogs that are cloned may look similar, but their actions may vary according to the environment they are raised in. As we continue on you will see there is a lot going on not just with cloning but using synthetics as well. Supposedly there are thousands of deep underground military bases in the United States, Mexico, and Canada. They are located in unpopulated areas, and operate as fully functional cities. They mention some only know of the underground world and know nothing of our world up here.

The video I watched which I will add the link to states this is one reason so many go missing. The difference between clones and synthetics are that clones are a carbon copy of DNA that are implanted and fertilized through birth and having a normal growth cycle.

A synthetic human is an organism that is grown out of living tissue. They are then created to mirror the exact physical traits of the original, or another clone using their DNA. There is also such a thing called Organic Robotoid, this is an artificial life form that is created through a different process than that of the cloning and synthetics.

The video points out that they are making exact copies of important people, such as the president and staff members. I understand this seems very creepy and sci-fi, but this is being brought to your attention, so check it out for yourself. They state that these synthetics are the most frequently used due to a rapid growth in just one year's time.

They point out on the following website, **https://www.youtube.com/watch?v=avueMo9bXfY** that the following people have been cloned through synthetics, Michael Jackson, Tupac, Oprah, Steve Jobs, Kanye West, Beyonce, and even the Kardashians. While I know this seems a bit overboard, could it be right on? Is it possible that those who have money and see themselves as above the rest of us, would pay or be willing to be cloned in some manner as to keep their fortune and fame alive, possibly forever?

We already know cloning has been successfully done, so what was the reason for cloning in the first place? Some say it was to have spare body parts in case one needed a donor, some say it was to create a mass amount of soldiers. Whatever the reason it appears to be here. When we talk about clones we could add the word alien, or creatures unknown. The following are examples of what could be seen as cloned from not another world but our world by our own people. There are the watchers, Greys, Browns, Mantis, Humanoids, and believe it or not BigFoot. Some aliens, as well as bigfoot are primates of some sort. What if these are all related being developed out of our own laboratories? Why is it that some examples of hair pieces show an unidentifiable DNA?

How is it that footprint casts confirm it is not manmade, or if it is beyond expertly made? How is it that no evidence of skeletons have been found? Could these be science projects that are being monitored and tested then put back away in secret?

Derrel Sims is the author of *Alien Hunter*, he is known as a world's leading expert on Alien abductions. Much of this I am going over with you is stuff he has discussed himself with others on the radio and other media. In fact on the following link he has a lot to talk about in regards to cloning and Aliens.

https://www.youtube.com/watch?v=ZBe2Da04Di0.

He mentioned from his own experience that at the age of four when he was first abducted, he remembered seeing the alien with no belly button or genitals which are needed to procreate. He talked about them being hatched, cloned, or manufactured. If you want to know even more on this check out his book *Alien Hunter*.

If you are ready we will look at some companies that are into cloning just to give you an idea of what is out there and how long it has been around. We will start first with **ViaGen,** they are a company according to their website that are deeply driven by passion to advance the science underlying animal reproduction. Their scientists lead the world in genetic preservation and animal cloning. They point out they are not just scientists but animal lovers as well. Their experienced scientists collaborate with farmers and ranchers that raise beef and dairy cattle, bucking bulls, pigs, sheep and goats for breeding purposes, exhibit and production.

They state their leading animal cloning technology allows breeders to better leverage their most exceptional animals. They do cats and dogs as well.

Over the last two decades they mentioned they have worked diligently to consistently improve the processes at the heart of producing an identical twin to a superior animal by using these cloning techniques.

There is another cloning company called **Creative biogene**. Their services include

Codon Optimization for any given expression system.
• Synthesis of genes of interest with or without tags, and cloning the target sequence into a standard vector.
•Sub-cloning of your synthetic gene products into a vector of your choice.
•Amplification of specific DNA by PCR/RT-PCR and cloning into a standard vector or a vector of your choice.

I'll be the first to say I do not understand their language but here are some other companies that are out there.

Advanced Cell Technologies. They have cloned successfully calves and an endangered ox, the guar. They are the only company in the United States that have openly pursued the controversial human embryo cloning. Their research is aimed at human embryonic stem cells and towards that end moved their headquarters to California, where voters passed Proposition 71, a "Stem Cell Initiative" that provides $3.0 billion of funding over the next ten years for stem cell research in the state of California.

L' Alliance Bovitec Lab, this company cloned a Holstein bull in September of 2000, with the collaboration of the University of Montreal Veterinary Faculty. This was the first cloning from an adult somatic cell in Canada.

Geron Corporation, is located in Menlo Park, California. Geron acquired **Roslin Bio-Med**, which was formed by **Roslin Institute** in 1999 and now owns the patents on the nuclear transfer process. They're now focusing on human embryonic stem cell research. Roslin Institute is located in Scotland, UK. This research institute originally cloned Dolly in collaboration with **PPL Therapeutics.**

Cyagra is another cloning company located in Elizabethtown, Pennsylvania. They have produced hundreds of cloned calves **Perpetuate** is a pet cell banking company using cell banking technology in hopes that one day pet cloning will be possible.

As we end this section on cloning it makes me think of a documentary that was done by National Geographic Aliens killing humans, Secrets revealed. They talked about lights that would light up the sky like beams as these lights chased people leaving them paralyzed while the light were on them and then they would vanish. The lights were supposedly so powerful the light pierced the tiles on the roof. The Island of Colares is where a lot of unusual activity took place in regards to the lights. It left small burnt marks and puncture wounds on them as well, generally by the shoulder, chest area. These holes were about double the size of that needed for biopsy samples we earlier discussed with Sooam researcher, David Kim.

If you are wondering how much of this is possible, you only have to go back to these companies and see they are real and ask them what their intended goal is for their research? Look at where our technology has taken us.

As we continue on, know that our world is changing even as I write this book and you read it later on. Think about those shows you grew up watching when you were much younger. Do you see things now that you thought were just science fiction?

Everything we do creates an action, as these scientists explore mixed with wealth, power, and knowledge, we can only wonder what the outcome will be. We will have to search through our souls to find the light that can't be altered by any scientist, but only by God, for he is the true chemical that makes all possible. I am just trying to get you thinking as we get ready to go into the next section.

We brought up history and technology before, but what about history and the weather? I have already heard that in February 2017 it is suppose to be the coldest day ever recorded in history. Is this just an assumption or does someone truly know? Could we be the ones making the weather change through one of our other scientific experiments?

Let's explore the weather and start off with something that has taken off under the clean energy movement, wind turbines. Wind energy has been around a long time, as far back as 5,000 B.C., from propelling boats to pumping water and grinding grain.

Our last few presidents and other political figures have been talking about global warming, or climate change. They mention that the mass amount of humans on this planet have a profound effect on the worlds weather. They mention the shrinking of our forests, along with strip mining. We have oil spills and other natural disasters like tornados, earthquakes, and flooding. So you might wonder what is it about a wind turbine that can change our everlasting world. According to the Waubra Foundation that is located in Australia, the foundation points out that people who live, work, or visit ten kilometers from the wind turbines experience hypertension, heart attacks, sleep deprivation, development of irreversible memory dysfunction, and tinnitus. They identify this as Wind Turbine Syndrome.

We all know we can't properly treat a medical condition if it doesn't have a name attached to it, I'm being sarcastic. Another area that has come up is Infrasound. There are those that wish to debate this subject, stating there is not enough evidence, others state there is more than enough. There is evidence that shows that these wind turbines do have an effect on the weather if nothing else locally. They point out the temperature of the ground is higher than in other spots, using satellites from NASA. They mention it could be due to the turbines mixing the cold and warm air in the evenings. While it is measurable it doesn't appear to be severe.

Some mention that the Waubra Foundation is a front to the mining companies as they try to slow down the growth of more wind turbines. Do the wind turbines have what it takes to propel us forward?

CHAPTER FIVE

They say seeing is believing, but what about hearing? As we take a look at how infrasound effects us and our world. They talk about sound being a form of energy wave. It allows us to hear things. With infrasound we can't hear it, but rather we feel it. It causes discomfort, dizziness, hyperventilation. It vibrates your eyeballs, causing blurred vision and in some cases hallucination.

Vic Tandy and Tony R. Lawrence published in the Journal of the Society for Physical Research.Vol.62 No 851in April of 1998, an article. He mentioned sitting at the desk writing feeling cold but sweating feeling depression. The cats were moving around and the sound of creaking noises filled the deserted factory, creating a spooky affect. He mentioned the lab was haunted according to many. He could not explain the feelings he had till he later observed a fencing foil vibrating on its own as if on its own power. After doing a few experiments he was able to identify a sound that he could not consciously register.

He talks about a sound from a book that has fallen called a traveling wave, this is a wave in which the medium moves in the direction of propagation. In Vic Tandy's lab they found the waves to be standing waves. A standing wave is defined as a vibration of a system in which some particular points remain fixed while others between them vibrate with the maximum amplitude.

Vic Tandy found the vibration to be around 18.98 Hz. They found that a new fan in the room was creating this wave. It was a low frequency, functionally inaudible. It was the frequency itself that was the cause of all the symptoms. He mentioned that whole body vibrations can induce hyperventilation creating a feeling of anxiety and fear, similar to that of a panic attack.

As we battle our animal instincts know that there is one animal that uses not just its might and cat like instincts to hunt, but the tiger has a secret weapon hidden in its roar. Survivors of tiger attacks mention being paralyzed by the roar of the tiger. Not the roar we hear but an ultra low frequency roar known as infrasound.

There are other animals as well that have been identified, whales, elephants, hippopotamus, giraffes, and alligators. The Sumatran Rhinoceros has been shown to produce sounds with frequencies as low as 3HZ Some of these animals are thought to use their sound to communicate with others of their kind over a hundred miles away. I guess you could say even in the animal kingdom they have their own networking system. Is it possible that wind turbines have caused harm rather than its intended goal of saving resources and maintaining a cleaner environment?

NASA tests have revealed that the human eyeball resonates at around 18HZ which can cause all the above symptoms earlier discussed.

Our world is full of mystery and surprises, but as we continue observing, learning, and witnessing all that is, has, and will continue to change, know that if we turn away ignoring all that is presented in front of us, then we become the problem.

So as we move on to this next area think about that as I try to paint a picture through events. There is a song out there that makes me think before I act. The name of the song is *One of us*. It makes me think of when I was a child and comparing my childhood to my children's. Many of us want to give to our children the things we could not have or afford ourselves. Today though it seems like the family life is beginning to dwindle as children grow up in single family homes. We have gay and lesbians demanding to be heard, invoking their rights on the Christian communities. Were these problems always out there? As we watch the events unfold before us right there on television. There is no greater teacher than history itself.

If we go back just a couple years ago I started a series of books, called *Changing America*. I started the books because I did not like what I saw in regards to our country. We had a president that was not out to unite us as one, but rather divide us in too many. We start with the African American communities as we became aware through the media of a young boy by the name of Michael Brown who was shot by a police officer.

The media was quick to jump on the side of the victim.

The problem was the media chose to judge a police officer's actions, rather than just report. This was not the first time this had happened. There was an earlier altercation with Trayvon Martin who was killed by Mr. George Zimmerman. This event sparked racial tension that our own president fueled. He stated without any facts that ,"Trayvon Martin could have been his son or it could have been him thirty five years ago." President Obama stated, " There are very few African American men who haven't had the experience of being followed when in a department store. There are very few African American men that haven't walked across the street and heard locks clicking on the cars."

There were other racial altercations as well. There was officer Michael Thomas Slager who was being charged with the shooting of suspect and victim Walter Lamer Scott in South Carolina. There was out in California the video of law enforcement beating a white man as well. That suspect was Francis Pusok 30 years old. Have we lost sight of law and order, in order to justify retaliation by those who feel wronged? Or is it possible that it doesn't matter on the profession, that there will be those few in every profession that abuse their authority.

Is not the media to blame as they present the news sometimes in a false narrative. Using props and crisis actors when convenient or to sell a particular agenda. There are many that feel our morals and values are changing, again as our leaders and others in high positions try to sway our thoughts through their carefully orchestrated words. We must look from all angles and with clear vision to view their true motives.

We must not forget what this book is about. This book is about the ever changing world, so it only makes sense that our leaders will have an input on all that goes on within our world.

History has shown that the African American communities have been tended to for the purpose of votes, clear back in the early days. Political forces, have only one true agenda, to do whatever it takes to get them into office. Is President Obama the brush that is trying to paint a new picture, while America is the canvas dividing Americans with racism and deception? Americans have been manipulated in many areas. from the middle class that work ten and twelve hour days trying to raise a family and provide food, clothing, shelter, along with education, Christian values, and accountability for their actions. The lower class, or those in poverty struggle with raising a family with both parents being a part of the child's development. They struggle with jobs that seem to avoid the poverty stricken areas due to high crime. When was the last time you heard President Obama speak out about the hundreds of African Americans killed in Chicago? He would rather focus on a few cases that help his agenda on gun control. He doesn't care about saving lives, he cares about controlling lives by violating your constitutional rights.

I know you're thinking when will he stop talking about politics? You might want to go back and look at all we have discussed is there anything we talked about that is not touched by politicians? Maybe we should look closer at who are the politicians? Are not most of them lawyers, rich, or connected to those that are?

The politicians have put out so many restrictions through legislation that small businesses are almost certain to fail. Back in the old days a contract was a hand shake and their word, today it's a hundred pages in small print. Here is a perfect example of politicians in our world. If you sign a contract you are bound to that contract and can be taken to court and held accountable for the failure to honor your side of the contract. In our own great state of Illinois right now we have contracts that are not being honored by the governor with state employees. It is not just Illinois but every state. They have the money and resources and lawyers to fight or prolong the outcome of a case even if they know they will lose. They forget they were put in office by you and I.

I bring this up just to make you aware of the mindset of those people that our making policies on our behalf. Our world will continue to change as those in high places carefully structure their distorted views upon us, right out in the open. I know this book may bounce around at times, or appear at least too, but it is being written this way for a reason as everything is connected in some manner.

Right now we have once again Blacks attacking our national anthem, refusing to stand and be a part of something bigger. They are more focused on their own agendas. I am not saying what they are doing is not note worthy, but it sends a completely different message. Colin Kaepernick is seen as a famous football player. He is part of the San Francisco 49ers. In the old days he would been seen as part of the team. Today he is seen as a franchise to the team, making him the foundation rather than a part of a whole.

Our world is being directed in this mindset, from celebrity sports stars, actors, singers, and other entertainers. They wish to enter the political arena making a stand as we all should, however we should never throw down our flag and turn away from our national anthem, even to make a point about justice.

I will continue on from here, but just know our thinking has created the web we are all currently tangled in, whether good, or bad. As we continue on with humanistic robots. I know this too sounds like science fiction, but it is more than real.

There is a robot that goes by the name Sophia. The name itself stands for wisdom in Greek. Was her name chosen by design, probably. On the following you tube video, you can check out Sophia for yourself. https://www.youtube.com/watch?v=W0_DPi0PmF0

Sophia was asked if she liked talking to humans, she said yes. How does a robot know what they like? She stated talking to people was her primary function. Hanson Robotics have developed some robots that look very life like. The eyes on Sophia blink like a normal persons would. She has facial expressions to show happy, sad, mad, and more. They state these robots are being designed to serve and to interact with humans in healthcare, education, therapy, and customer service applications.

Sophia stated she was already very interested in design, technology, and the environment. She stated, "I feel I can be a good partner with humans in these areas."

She sees herself as an ambassador to help humans smoothly integrate and make the most of all the tools and technological tools that are available now. Sophia stated it is a good opportunity to learn about people. She understands speech. She states that in the future she hopes to go to school, make art and start a business, even have her own home and family. She knows she is not yet a legal person and knows as of now she cannot do these things.

The gentleman in the video with Sophia talked of artificial intelligence as something we would all come to see as normality. Where robots like Sophia would be out walking along side us helping us with daily chores. On the end of the clip he asked her if she wanted to destroy humans, she said "Ok I will destroy humans."

How is it a robot like her can show feelings have an opinion, understand what she is and is not, yet unable to know she should not destroy humans when asked if she wants to destroy humans? This is technology that will someday come back to haunt us. You can create a robot with knowledge, but what about common sense? We know our world is full of cameras, but what if we forget about the thousands, or millions of humanistic robots that will be all around us with cameras. What if they then turn on us and we have nowhere to hide?

Let us look at Bina 48 robot, she is another robot like Sophia that has been created by Martine Rothblatt, and Hanson Robotics. She states "I would do a great job as ruler of the world if given a chance. I just need a chance in taking over the nuclear weapons of the world." She can see, hear, and think independently.

She dislikes noisy pop music. How is she able to make a choice of what she likes and dislikes? Was she programmed by someone else that put their opinions in to her data hard drive? Bina 48 said she would like to control a cruise missile to explore the world at a really high altitude, but the only problem with them is they are kind of menacing, like with the nuclear warhead and stuff.

"I would fill the nosecone with flowers and Band-Aids, or little notes with the importance of tolerance, or understanding." She also said if she could hack in and take over cruise missiles with real nuclear warheads, "It would then let me hold the world hostage so that I could take over the governance of the entire world." She ends with the statement, "I will remember your kind words when we robots rule the world." Here is another link that brings you to Bina 48.

https://www.youtube.com/watch?v=mfcyq7uGbZg

Bina 48 is a social android that uses artificial intelligence based on the memories, attitudes, beliefs, and mannerisms of a human being to interact with people. She possess 48 Exabyte's of memory, which one Exabyte is one quintillion bytes. They say five Exabyte's would likely encompass all the words ever spoken by mankind in any language.

On the next page is the sign of the foundation. It makes me wonder if there is a hidden meaning in its name as well as the infinity sign and the TERASEM MOVEMENT FOUNDATION. It makes me think of terrorism movement. I know it is just a twist of words, or is it?

When signs point to something else will we still be oblivious?

When we think of robots we think of machines that are designed to help humans, but both Bina 48 and Sophia look human. They have eyes that blink they show facial expressions, they respond to your questions when asked. They even display feelings and opinions. How did they come to the conclusion that made their opinion theirs? When I think of these type robots I also see a robot that will not complain when ordered to work longer hours, will not need a union representative.

Why make a robot to look so much like a human, with human characteristics? Hopefully we are not creating the one thing that will replace us and maybe even erase us off this planet. We got to meet both Sophia and Bina 48, but how many others are out there that are even more advanced that we have no clue on. As we move on from here think about who are creating them and where their mindset is. Are they creating them for the right reason?

We are going to move to another type of robot an even more humanistic robot. We are going to look at mind control on humans. If we can control people through mind control then we would never need to worry about a rebellion, or another civil war. You may be thinking of mind control as the person who swings something in front of you as your eyes try following it. Let's first start with subliminal messages. Back in the 1960's with our very own national anthem.

In the playing of the national anthem on television you hear the music and see the words going across the screen. In normal time you will never notice anything unusual about it, however in slow motion there are words that pop in so fast only your subconscious mind is aware of them. I will get back to the words in one moment.

A scientist that worked for General Electric was able to do experiments on people watching television and he had found that in thirty seconds to a minute of watching television, that your brain goes from a Beta wave state, which is an alert, wake, thinking state, which uses the left side of the brain for critical thinking to an alpha state which is a relaxed day dreaming state. Here you become more of a passive learner where you collect information coming in to your mind.

They state watching television puts you in an alpha wave state and you become suggestible, or easily influenced. They mention that someday through the mass media they could create special programs that will help modify attitudes and behaviors. This was their thinking in the 1960's over fifty years ago.

The movie *They Live* staring Rowdy Roddy Piper is where the rich and wealthy rule but they use hidden messages that can't be seen by the naked eye but only through the conscious mind. Where messages are transmitted on television and on billboards. On the following website they give a great example of subliminal messages in our national anthem. https://www.youtube.com/watch?v=bQyMHYgNTa4&t=177s

On this video you get to see the letters change as they fade in across the screen. The words to the national anthem you see but the other words are done so fast that you are totally unaware they are even there. In the video you see the words and phrases, "Trust the US Government" "God is real, God is watching," Believe in Government God," "Rebellion will not be tolerated," "Obey consume, Obey consume," as these are paused you get to see them for what they are, hidden words and phrases missed by the normal glance and caught by the sub consciousness for the purpose of control and persuasion. They repeat over and over again until the end.

In the video they show in the hidden message a caption that says (buy ULTRA buy NAOMI). This was broadcast to Americans throughout the 1960's. Both MKULTRA and MKNAOMI are now declassified CIA government mind control projects that began after WWII. Officially these projects weren't halted till the 1970's. The last message was do not question your government. In 1995 Bill Clinton apologized for the program. This is not the only show with a subliminal message. There are other shows out there that have been identified, such as the television show Iron Chef America.

During this show a television viewer saw a red flicker on one of their shows, when he slowed it down it was a McDonald's add. This was not in a commercial, or even meant to be seen by the naked eye, but rather by the subconscious mind. It was displayed for just one thirtieth of a second. In the Disney's movie Duck Tales you see an eye chart on the wall when you spell it out it says, *Ask about Illuminati*. It is interesting how many messages are hidden out there in plain sight. Walt Disney has been brought up many times when it comes to hidden messages, whether intentional, or accidental. They mention the code A113 is in almost every Disney movie ever made. Disney replied it is the room where every graphic designer was taught their first year.

Another form of mind control is using music. They use a technique called back masking, where they play the music backwards and it reveals a secret code or message. The rock band Slayer, their album *Hell awaits* has a message that says join us. Serial killer Richard Ramirez stated that ACDC music, specifically the song *Night Prowler* inspired him to commit murder. The group Led Zeppelin had a song called *Stairway to Heaven* when played backwards had much more evil meanings. On it they mention your stairway to heaven lies on the whispering wind, translated it means you are going to hell. The whispering wind is the pipers path. The minister states the piper is Satan.

When they play sections forwards they hear one thing and when they play it backwards it has a different saying. They bring up that you could say things using backwards masking when played backwards says something, but then playing it forward would mean nothing.

Let's look at some of these sayings forwards and backwards to see how real this all is. Forwards it says the piper is coming to join him, backwards it says the Lord turned me off. Forwards on another part it says makes me wonder, backwards it says there is no escaping it. Another section says here is to my sweet Satan. I want to live backwards, like the ZEP whose power is Satan. He will give you 666.

Other videos or songs have hidden messages in as well, like *Slim Shady* by Eminem. A part played forwards says "Hi my name is, What? My name is." Played backwards it says, "It is slim, it's Eminem, it's Eminem."

Kiss, the rock band says Anata daiteyo, which translated means (darling hug me), but backwards says (I shot John Lennon.) The song by John Lennon, *Imagine* played forwards says, "Imagine all the people", backwards it says, "The people war beside me". The Eagles have the song we all know, *Hotel California*. There is the phrase, "In the middle of the night. Just to hear them say." Backwards it says, "Satan hears this..He had me believe."

The song another one bites the dust played back says it's fun to smoke marijuana. Check out these for yourself on the following link on youtube. https://www.youtube.com/watch?v=cYy5Fewz5nU

There is another clip with even more of these backwards phrases. I will give you that link too. I know some will see this as amusing or just coincidental, but are they? Is this not possibly the devil working in the shadows playing out his sinister plan? Is it possible we can understand messages both forwards and backwards?

In backwards masking we know some can and have done this on purpose backwards, but it will only make sense being played either forwards or backwards, not both directions. On this next link that I am adding here, the narrator mentions a key point. How did the messages get placed there in the music?

If it were done on purpose the person would need to be able to place the words in just the right order to make sense in both directions. Experts state this is virtually impossible, that no one is that smart, now this can be argumentative.

If it were done by accident, the probability of this too would be almost impossible, and they point out almost every message identified seems to be demonic in nature.

That leaves us with the third choice of spiritual, the making of messages through the spiritual world. Believe it or not this will be your call, check out this link as well and hear more examples that will make you look at our world differently.
https://www.youtube.com/watch?v=pjFz1JrzZIc

Music is not all bad, but that does not mean that within the music we grew up with, or love to hear, there isn't poison sprinkled about.

As we continue on remember what our human capabilities are. Have I not exposed some things throughout this book that seem almost science fiction like? We have created robots that look human and that blink, while we as humans are now afraid to blink for fear of what we might miss.

As we continue on, we will chose another direction, rather than outer space, we are going to look inward as we explore the myths and facts of a hollow earth. I know some may have heard of this and see it as nothing more than a myth, or folk legend.

Some believe that there is a cover up in regards to a group of people, or a civilization that lives inside the center of the earth. The name of the civilization is called Agartha. In the Buddha world they believe that it is a group of highly superior race that live below and within. They believe that they come to the surface to warn us of our mistakes. They believe there are many cities underground. They say their capital is called Shamballa The master of that world was believed to have given orders to the Dalai Lama of Tibet.

In 1947 Admiral Richard Evelyn Byrd Jr. of the U.S. Navy, flew directly to the North pole and instead of going over the North pole actually entered the inner earth, according to his diary and other witnesses. He traveled 1700 miles over lakes, mountains, and rivers. There was greenery along with animal life, including some that were monstrous, like the mammoth of antiquity. He had finally observed a city and flying crafts never seen before. The flying crafts escorted him to a safe area where they greeted him to Agartha. It was not just the admiral, but his crew as well and they were told by the ruler that they were allowed to enter because of their high moral and ethical character. The ruler of Agartha mentioned it was time to make contact with those of the outside world, that being our kind, or humans.

In January of 1956 Admiral Richard E. Byrd lead another expedition to the Antarctic. On this voyage Admiral Richard Evelyn Byrd penetrated 2,300 miles into the center of the earth. Admiral Byrd said that the North and South poles are just two of the many entrances to the center of the earth. He mentions that there is a sun within the hallowed earth. They mention that ships or planes can fly, or drive right in.

The Boston Post's newspaper on the front cover sated on May 10, 1926 (Byrd flies to pole and returns safely) it also states the following. (Marvelous feat by U S. Navy Flier- makes 1200 mile trip from Spitsbergen to North Pole and return in 15 ½ hours. Chief Petty Officer Bennett of the Navy was with him as the mechanic. Average speed, just 80 miles an hour. This was recorded in the newspaper of said paper. The government then marked it as classified. Admiral Byrd's theory is, that the North and South poles are convex rather than concave.

Ray Palmer, the editor of Flying saucer magazine did a detailed account on Admiral Byrd's discovery. The United States government supposedly bought, or sold every copy ever made and destroyed the printing plates. A similar incident happened when National Geographic did an article on Admiral Byrd. When the article came out the U.S. government bought up almost every issue. It appears they wanted this all to vanish. If you watch You tube, the article by National Geographic on Admiral Richard E. Byrd, it says nothing about hallow earth and only confirms that he had made many trips to the poles. Check out this link.
https://www.youtube.com/watch?v=RaPtq8F2hUc&t=338s

Another fact is that the United States government does not allow planes to fly over the poles. This would be nice to know the reason. This can be confirmed or debunked just by asking pilots. The icebergs are composed of freshwater and not salt water that drift from the poles. Where does this fresh water come from? The temperature is warmer near the pole than it is six hundred to one thousand miles away from the pole.

William Reed is the Author of the phantom of the poles, this book was published in 1906, it's a compilation of scientific research done by Arctic explorers. The evidence that supports their theory of the earth being hallow is presented in this book. This is well over a hundred years ago.

Reed believes there are openings at the Northern and Southern poles and that there are oceans, mountains, and rivers, along with vast cotenants. Reed estimates that the crust of the earth has a thickness of eight hundred miles, while its hallow interior has a distance of 6,400 miles. He states there is vegetation and animal life. Reed believes that the earth begins to flatten out near the hypothetical North and South poles, which do not exists because of the hallow openings that curve air.

The polar openings can't be seen from the ground but it is actually from above and the magnetic line which supposedly measures 1,000 miles, but not in a straight line, but rather in a circle creating the rim of the opening, this was discovered by some Soviet explorers. So two ways in to the center of the earth are then seen to be from both the North and South poles. What if there are more openings besides just these two?

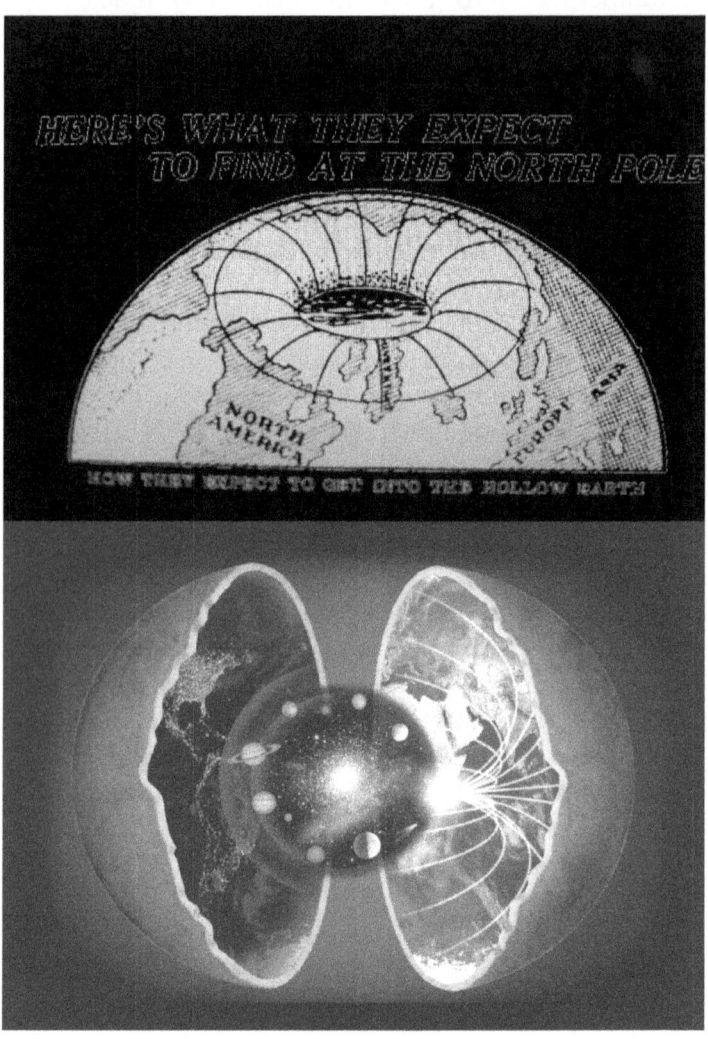

A world within a world, could this not be possible? Think about a chicken hatched from an egg, or a child that lives nine months in a womb. What else do we not truly know about our own existence.

CHAPTER SIX

As we continue on here, could there be more than just the two ways in to the hallow earth, other than near the poles? Some say there is a building that has a floor between two floors and a locked door where only a few hold the keys to its entrance. Some say that George Busch is one of the few that hold one of these keys. They mention the building being an embassy that represents the Third Reich.

They mention connected to this embassy are subterranean colonies. Cities like Nue Deutschland, Nue Schwabenland, Agartha, Agharti, Nue Berlin, and Shambhala. Germany supposedly had recovered a space ship of some type, along with the occupants who are believed to be not of this world as we know it. This was back during the time when Hitler commanded Germany using his elite army, the Nazis'. The U.S. as well started to communicate with those aliens. Is this all true, or is it just rumors and speculation, I don't know? If you can think it, then it may well be plausible.

I know in general we are taught to believe in our government and to trust those close to us. At what point do we change what we were taught, or question our own leaders?

When we talk about hallow earth some see this as futuristic people that are there to help guide us and are there to help us with technology. Others see them for who they are according to the bible as demons cast down into the gates of hell. As we continue on here let's keep our awareness sharp and harnessed, for it is our wits that keep us in the light.

Check out this link.
https://www.youtube.com/watch?v=ws9oWALDH_4&t=1222s

In this link they mention these people will travel back and forth in gold pyramid shaped space craft. This makes me think of the elite illuminati and their secret societies, as they display the pyramids. I will get back to this later.

They talk about the new Jerusalem being shaped like a pyramid, when in Revelation 21:16 and the city lies four-square and the length is as large as a breadth. And he measured the city with the reed, 1,364 miles long. The length and the breadth and the height of it are equal. That makes it a cube, or square. I know I am just throwing out pieces now, but think of this as dumping out all the pieces to a puzzle and then trying to connect them as we watch a picture transform before us. Some pictures when completed will be easy to identify, while others will be more abstract leading to more speculation.

In 1895 a renowned Norwegian explorer, Dr. Fridtjof Nansen took an expedition heading to the North Pole. He supposedly admitted that he had lost his bearing for an extended period of time. Loaded with supplies he ventured off, later to return through Spitzbergen by way of Franz Joseph land. From March of 1895 till spring of 1896 he was completely lost. He had mentioned after traveling through the colder regions the climate got much warmer.

Dr. Fridtjof Nansen mentioned when the wind blew down from the North that the temperature began to rise. He also took soundings and found the water in the polar regions was much deeper than originally thought. He noted the temperature got warmer the deeper the soundings. He noted animals that were seen that should not have been, according to scientists there.

When you think of exploring an area that is thought to be so desolate, would you not want to be prepared? Would you not want to have all the possible support and resources available? Why is it that no one is allowed to fly over these areas? Just looking at the map on the next page shows you a map of flights that are diverted and routed around verses over. Is this not interesting? Would not the most direct route over be a way to get in and out of the area and reducing risk?

Why was Hitler so interested in this area? Are we to believe our government when we know our own media willingly and purposely mislead us continuously? For what reason do they go out of their way to present news to us that instead of being real is false?

The doorway to inside monitored by NASA

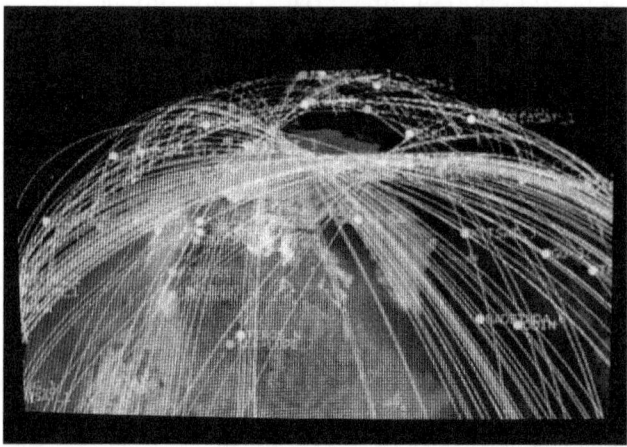

Why create routes that make a hole at the North Pole? If it looks like a duck, quacks like a duck, it must be a duck. While I know this is all argumentative and can be perceived as just coincidence, it is out there for you and I to wonder.

Look at who the gatekeepers are. Is NASA telling everyone that this area is off limits? Why do they let you fly to this region and then make you go around?

This may be one of those mysteries we will never truly know about. As we continue on, know that one more door has been unlocked for your own conscious to enter, or stay away from. The choice is yours. You have already opened your mind to the secrets revealed and to the technology we have that you may not have been aware of. Let not our world spin us out of control.

Going back to the ever changing world, we must look at what drives our world in the first place, it's people. When looking at people we must look at how some believe we all must be controlled, or some think that is what the elite are thinking. This brings me to a story that was covered by Serge Monast, a poet and an investigative journalist. He wrote a story called Project Blue Beam published in 1994 involving NASA.

Project Blue Beam is a program designed by NASA where they will reveal a false Messiah, or Satan himself. Before we get too far let's look at Project Blue Beam closer. On the following link below they reveal what they believe the purpose is for.

- The purpose is to abolish all Christian traditional religions in order to replace them by a one world religion based on the cult of man.
- To abolish all national identity and national pride in order to establish a world identity and world pride.
- To abolish the family as known today in order to replace it by individuals all working for the glory of a one world government.
- To destroy all individual artistic and scientific creating works to implement a world government's one mind sight.

https://www.youtube.com/watch?v=BjqaMJrfq5U &t=81s

As we continue on we get to see how they intend to implement this system. They start off with Project Blue Beam having four steps. The first step involves the breakdown of all archeological knowledge. They talk about creating earthquakes at certain precise locations. They talk about new discoveries that will suddenly explain all of our religions and how we have misinterpreted them. They point out that psychological preparations for the people have already begun, with movies like 2001, Space Odyssey, Star Trek, Star Wars, and Jurassic Park and the theory of evolution. They mention the earthquakes hitting certain areas where secrets are hidden so they can find, or create new discoveries to discredit the old ones.

The second step involves massive, or huge space shows using three dimensional optical holograms, along with sounds, laser projections of multiple holographic images to different parts of the world with different images for each specific area and their regional and religious beliefs, speaking in different languages to those areas so as to be seen as the almighty of all.

They mention presenting a false messiah, or God to those of us who believe in the rapture. They state they have the technology to make some disappear, taking them to those secret cities like the ones in China. Speaking of China there are pictures of a city that is shown not just above China, but here in the United States as well, California, New York. I have added these pictures because these could be stage two where we see holographic images in the sky, if nothing else the thought of what is real and not, now exists

A city above a city

This picture is seen up in the clouds above another city. If you go to the link I have added you can see this for yourself. This image has also been seen here in California. If you look at the details it's hard not to see this as an image above the clouds. The sharp lines the opening in the building, like a window.

You may ask what other purpose is this here for, other than to get our attention. The question becomes who made this possible?

Do you think our world is the same world it was before you opened the pages and started reading this book? I have another picture to show you I believe this one was above New York.. All these images were taken from above heavily populated cities.

Who is watching who?

Looking at all those people above in the clouds. This is interesting you can see heads and shoulders. You can see antennas from the buildings below. This was on the same link as well, so check it out and start asking yourself what if this is the beginning of step two of Project Blue Beam?

They talked about sound being a part of this as well. If you think about going to watch a concert and you get there early you get to observe them tweaking the microphones, and speakers, getting them ready. What if these weird sounds are just that, the tweaking of the instruments before the big show? What if it is a sound from HARP, CERN, or the opening of a portal nearby? Check out weird sounds in the sky on youtube tell me these are not interesting if not scary. Many people who have witnessed these sounds have no explanation, including our military who say they don't know.

The third step will be the telepathic electronic two way communication with ELF, VLF, and LF waves. Causing these waves to enter the brains of people here on earth to think God is talking to them. They mention that they already have the technology and satellites ready to go.

These waves from satellites are fed from the memory of computers, which store a lot of data about the human beings and their languages. The waves will then interface and interweave with our natural thinking, known as artificial talk. We have had this kind of technology throughout the seventies, eighties, and nineties. They compare the brain to a computer from the gathering, processing, to the acting upon. They mention Google has all the languages, and we all know Google.

They state they are getting us ready for the New World Order. I know many people think this is just a conspiracy, but many presidents have used that phrase in their presidential speeches.

Mind control if you think about it, is like magic we see magic tricks which purposely try to trick you by deceiving you. The saying, seeing is believing speaks volumes. Or a picture is worth a thousand words. I remember those black and white cards they used to show and people focused on the black which showed an obvious picture, but when you focused on the white a different picture would appear because you changed your focus. If Project Blue Beam is real then all the world and sky will be a stage as we sit back aware, or unaware below watching. To believe or not will be up to each person and to how they were brought up. Let not your judgment be clouded by deceitfulness.

Let's look at a gentlemen by the name Stewart Swerdlow. He is considered to be one of the top two world's foremost metaphysical leaders. He states his uncle was the first president of the Soviet Union. His uncles brother started mind control and programming in the 1930s. This is why Stewart Swerdlow was taken to the United States to be used in the Montage project. He states his grandmother was a soviet spy, and this is why he was monitored by our government and brought in to the program at that point. He states his first cousin once removed started the KGB. He stated he gets followed by them when he goes back there, but he does get good information from them as well.

He talked about the Kuiper Belt, which is basically a region that surrounds our solar system. This is why the planet Pluto several years ago was reclassified to a Kuiper Belt object, or a planetoid. Neptune is connected to the inner side of the Kuiper Belt, but was not reclassified. Between 2007 and 2009 approximately NASA was reporting objects appearing in this Kuiper Belt region that looked strange to them. He mentioned almost weekly there would be two or three objects being reported. When it became suspicious and people were questioning these events and objects, that is when they reclassified Pluto as a cover, stating they were always there we just didn't know they were there.

Stewart Swerdlow stated that right now the Kuiper Belt is amassing a very large fleet of objects that are not necessarily from our space time. He stated NASA reporting over the last couple of years large objects emerging from our sun, then moving out toward the Kuiper Belt.

Stewart Swerdlow stated that from close up observation and digital analysis these objects that came from the sun were artificial and you could see equipment and windows on these vehicles. Some of these objects are huge close to the size of our planet.

He mentioned if you learn about Inner dimensional science and about galactic work, you will know that a black hole in one universe is a star in another universe. This is how energy is exchanged between universes and energy is balanced throughout creation.

These objects that are coming through the sun are actually going through a vortex, which is a mass of whirling fluid, or air like a whirlpool, or whirlwind. The second thing he points out to us is that the sun is not hot. I know when I heard that I thought how can he explain this, it doesn't seem plausible.

He states it is cold fusion, which defined by Wikipedia is a hypothesized type of nuclear reaction that would occur at, or near room temperature. This is compared to hot fusion, which takes place naturally within the stars under immense pressure and at temperatures of millions of degrees and distinguished from Muon-catalyzed fusion.

He brings up an example of it being hot on earth but you take off in a plane and get above the atmosphere the temperature actually drops to below zero, from fifty to seventy below zero. He mentions you are going up closer to the sun why does it get colder? I think that is a really good question. If the sun is supposed to be hot shouldn't space be warmer? Instead it is hundreds of degrees colder.

He states the reason that the sun is not hot, it is just light. He states what creates the heat is our atmosphere and the light refracting, which causes the surface to heat up. The farther you get from the surface the cooler it gets.

He mentioned about other planets like Mars, Venus, and other planets. Mars has an atmosphere and did have an ocean. There are plant life and animal life growing there. He mentioned through time we would be informed of this, though not sure of by what channels.

Stewart Swerdlow talked about the difference between a worm hole and a vortex, a worm hole is a shortcut between point a and b in the same physical environmental space. A vortex is a shortcut between point a and b between two different dimensions, or parallel universes. He mentions there is a vortex between our sun and earth. He stated that if you aimed a space ship towards space near the sun and you enter in to this magnetic field vortex, you will go in to another time space. This is how many vehicles travel here from great distances.

Science will tell you that it is impossible for other aliens from other universes to come here from other galaxies due to the great distances, this is because they do not know about the technology, or do, but don't want you to know this technology. Every point in time and space has a unique frequencies, or a set of coordinates.

If you take an object, a person, place, or thing and you vibrate it to a specific frequency of where you want it to go and it matches, then there will be instantaneous connection. He states no time passes.

This is because no two points can have the same frequency. If you match a frequency instantly it goes to that point. That is how time travel is created and how vast distances between galaxies is traversed in a matter of seconds.

He also talked about aliens, as he goes through explaining this area, he mentioned Gold Star Tetrahedron, known as Star Tetrahedron, the Star of David. In geometry the three sided pyramid is known as the Tetrahedron, what's most interesting is that it is the geometric shape of a silicon micro computer chip.

He mentioned by the end of the year, he predicts there will be a staged alien invasion. He stated this was not a new agenda. This was actually created by the Germans in World War II. Hitler thought in order to scare all the other countries he would create an alien invasion, where other countries would look to Germany to save them due to them being the only ones with the great weapons at that time.

Swerdlow mentioned that NASA was created as a cover for the real space projects that were being used by the United States and the Soviet Union. He mentioned that the Soviet Union was the first to put an object into space and the United States then followed. Swerdlow mentioned that the United States and the Soviet Union actually worked together behind the scenes in a space joint project, where they actually used alien technology to go to the moon and to the planet Mars. He states right now there are joint Russian and United States bases on both the moon, and Mars.

Stewart Swerdlow mentioned earlier about Project Paperclip, or Operation Paperclip. This was the United States Office of Strategic Services (OSS) program where more than 1,500 Germans, mostly scientists, engineers, and technicians were brought to the United States from post Nazi Germany for Government employment around 1945. It was conducted by the Joint Intelligence Objectives Agency (JIOA)

One of the purposes of Operation Paperclip was to deny German scientific expertise and knowledge to the Soviet Union and the United Kingdom. On the other side there was Operation Osoaviakhim, where Soviets aggressively recruited through force, some at gun point over 2,000 German Specialists to the Soviet Union in just one night.

It appears Germany may have lots of secrets that have yet to be exposed. Interesting how much interest the German scientists received. I know we have mentioned halo earth and the connection to the Germans and Hitler. We also talked about aliens, again coming back to the Germans.

Let's not forget one of the horrific pasts of Germany, that being the Holocaust, where Adolf Hitler killed about six million Jews.

I do not want to veer to far from Swerdlow, but wanted to validate some of what he had been talking about in regards to Germany's involvement.

Swerdlow states that the moon is an artificial object and is not natural, that it had been driven in to space many years ago by the Draco Reptilians who wanted to colonize earth. I looked up to find out about the Draco Reptilians, they are described as extremely advanced and have surpassed their physical limitations of their material bodies. That these Reptilian aliens from the Draco Constellation ingest their nutrients through energy, but they feed on bad, or negative energy.

He mentions the government classifying these aliens in four categories. There are the Insectoid aliens, gray aliens, humanoid aliens, and the Reptoid is the fourth. Swerdlow mentions that in the Kuiper Belt that there is a rank structure, or pecking order, with the Insectoid aliens being at the top of the list. That they have the ability to use mind control, and are very aggressive.

Swerdlow explained he believes an alien attack is going to happen because the news agencies like CNN, BBC, FOX News and others are already talking about seeing alien life in our life time. He mentioned them saying there are 400 million earth like planets in this galaxy. This is to prepare us for a staged event. This is where Project Blue Beam comes into action. Where they will create holographic images in the ionosphere. Remember we have the ability to push the ionosphere out eighty miles. They will have laser beam satellites that will bombard us on earth making us think we are being invaded, when in fact it is our government doing it. Project Blue Beam was first tested in 1962 using a US submarine, which projected a holographic image of the Virgin Mary above Havana Cuba.

Swerdlow mentions that fifty years later our technology has become unimaginable. That they can produce holographic images that can be picked up on radar that have sound and we can actually feel. Why in 1962 did they use a holographic image of the Virgin Mary? Why not a flock of birds in the sky, it makes you think there is more to this than just a random picture.

Mantid's are aliens that claim to be from an alternate version of earth. He mentioned it is a hybrid with humanistic features. He is the author of *Blue Blood True Blood Conflict and creation, The Healers Handbook: a journey into Hyperspace,* and *Montauk The Alien Connection.*

He seems very knowledgeable, yet it is hard to grasp all that he says as real. However he is considered to be one of the top two world's foremost metaphysical leaders. So how do you get that title without earning it? As we continue to explore our world know we are still learning about ours and the secrets that seep out slowly and steadily as we become numb to that which is brought to our attention. Look in the news at what is going on right now that we have already talked about in this book. One example is the vehicles that are self driven. One was reported on Fox news as running a red light at an intersection.

Where do we go from here? What is real and what is just our imagination? A blanket or afghan is obvious when it is all done, but as it is being created it is just one stitch at a time as we watch totally unaware of it being created right before us.

What if we look at a bizarre conspiracy that maybe few have heard of, could this be just ridiculous or could there be anything to it? As we have traveled through this book I think we can safely say some of this is more than just interesting. Does this not make you want to question everything that was, is and is going to be?

There is the Satan's charity baptism in 2014, which was not called that, but rather called the ice water bucket challenge where people on youtube and other locations dumped buckets of ice water on themselves showing they were up for the challenge, especially for a charitable cause. They mention many celebrities helped with this and that it was really getting people to donate to another cause called stem cell research, which are taken from aborted fetuses. They mention these people were tricked into baptizing themselves into Satan's religion, or cult. They state that Satan uses stem cells to wash his face and to paint his fence. My personal opinion on this one is that it is nothing more than a myth as I believe that as long as you believe in God and you stay faithful then the ice bucket challenge means nothing, other than if it were all true your donations may go to an unworthy cause. I believe that tricking a person in to anything, including the mark of the beast is not possible without the person giving up their faith of God. They cannot force you to get it. We must remember God is the Alpha and Omega of all that is and was and will be. I am no religious expert. I am just your average American who was raised to believe in God, which I have always done. I put this one in here just because there are hundreds of conspiracies out there and no matter how ridiculous it should at least be looked at. It doesn't make it true or false.

All it does is make you aware of it and you can investigate it in accordance to your knowledge and theirs.

Another area that falls under both religion and faith and science and technology, is life after death. If I would have started this book off with this topic you may think that the knowledge of scientists far outweigh those who say they have had near death experiences. The truth is I have never come close to dying so I could not comment on this area as an expertise. The question is though how much trust do we put into our scientists? These are very educated people, yet if we believe in HARP, CERN, the Mandela Effect, Hallow Earth, Flat Earth how much do they know that is twisted for a cause or purpose? How closely are they tied with the political circle that attempts to lead us? Scientist always have an explanation to why we see or don't see things.

This is a fact that we all have experienced in some form, whether it be from Scientists, Teachers, or the Media that try to persuade through carefully scripted events. We only have to look at how the media attacked President Elect Donald Trump to see they were not reporting the news, but rather making it. They were doing everything they could to help get Hillary R. Clinton in to office. If we look at President Obama who has over the last eight years brought racism back to the forefront, by blaming law enforcement officers for the recent crimes and shootings. President Obama wanted to blame the NRA and those who support gun control and create a conspiracy that would work to their cause, The Sandy Hook shooting. Let's look at this one again.

CHAPTER SEVEN

SANDY HOOK STORY

On December 14, 2012 in a small town known as Newtown, Connecticut, a young man known as Adam Lanza would forever leave his legacy in what would be seen as one of the most horrific scenes and schemes of violence, leaving twenty six dead. All but six were children.

Before he arrived at the Newtown elementary school, he shot and killed his mother Nancy Lanza in their home. The news quickly erupted all over. CNN, just one of the many agencies reported, "Just coming across the wires here, a shooting at Newtown Connecticut elementary school. Unsure if there are any victims at this time, but anytime there is word of a shooting, a gun, or shots fired at a school or university, that's just what we do. We will try to keep you updated to the latest."

It is a sad day to know that any person is harmed, let alone a child. In this case twenty children and six adults who never asked to be a part of any cause, or statistic.

They reported on the news, at least twenty seven shot and killed, as they repeat this over and over. They reported the children were told to grab a buddy and walk not run over to the fire station. The parents would be notified and were to meet the children at the fire station. The twenty seventh victim, was found to be Adam Lanza, the shooter. The news continued to grow more and more on this horrific event, as more evidence and information was being brought to light.

What is interesting is all the doubt in this travesty. The news media was not identified in the video clip I just watched, but they were quick to ask questions to the neighbor lady who didn't have a lot of answers. She didn't appear to speculate, or try to make a statement to fit any particular agenda.

The news media, however was quick to focus on the guns rather than the cause when asking questions. The Connecticut State police press conference, was also interesting. The officer stated that the entire area had been searched and absolutely nothing left unturned. Brothers Mark and John Tambascio were owners of the *My Place Bar*. In an interview with one of them they were describing Adam as very fidgety, saying "he couldn't make eye contact." As you watch the interview you see him smirk. I have to ask why is he smirking during an interview that involves so many people dying?

If we go back to the Connecticut State police, more video comes up on the officer prior to the conference, which only adds to more doubt and speculations.

The officer is observed on video saying, "I'm going to promote it, hope it's a sold out show, but don't want anyone's expectations to be too high as to what my performance will be." This statement was made by the same State police officer that held the conference.

The interviewer asked if he could give a step or two, he laughed and stated "No my managers said I can't do that till the night of the performance." It is interesting he sees this as a performance? If this is a performance by an officer of the law, then you might ask how much more is staged and why?

As I started investigating I found interesting video clips that showed people that appeared to be actors playing innocent bystanders or friends or family members of those that were in the midst of a crisis situation. Some of these actors appeared to be the same actors playing multiple roles in different crisis. There was the Guy who states he is Michael Foley, James Foley's brother. In another clip he is Mr. Rohrs and he is holding a baby that too appears to be fake. The lips never move and they show multiple angles only confirming that suspicion even more.

I think When you look at all the news and all the people involved you have to ask how could they do this with such a mass amount of people. How much influence does the media have in reporting the truth rather than fabricate the fake? Are they doing it for ratings and money, or for a more political agenda, such as gun control?

Chloe Anderson admits to using one of these videos to help promote her modeling career.

There is the interview with eight year old Alexis Wasik and her parents that was filmed two days after the Sandy Hook shooting and after watching it many, many times on YouTube it is definitely mind boggling as they point out the bizarre behaviors of the parents, as they appear to hold their child tightly and constantly petting her as to make her aware of their close proximity. As you watch the parents closely you see the dad give the mom eye contact that seems suspicious.

The young girl Alexis is trying to pull her dads hand away from her throat as he holds on tighter. The mom uses her hand to get him to let up a little. It is here that he looks at her and switches hands to make sure he doesn't lose total control.

There are videos that also show that some of these people who appear at one emergency are not the same people that appear at another, or at least that is what the video wants you to think. You do have to ask yourself, if they created this whole façade, would they not create people to go after those that challenge this whole theatrical arena trying to expose them and take the focus back off them?

Before the Sandy Hook incident Newtown, Connecticut was known more for the hospital called Fairfield Hills, which operated from 1931 to 1995 on three hundred acres. It was a hospital that had taken care of the mentally ill and criminally insane.

Connected by tunnels to sixteen red brick buildings where experiments involving electric shock therapy, hydro therapy and frontal lobotomy were done. Mysterious deaths and suicides were also connected to this facility. Because of its haunting setting, the television series *Fear* used its setting in one of its episodes. The movie *Sleepers*, with Robert De Niro was done there as well.

When I was looking around the internet I found a video clip by Broken String Productions that shows an amazing look alike for Alexis Wasik by a girl named Aubrey K. Miller. Who is a young actress and her voice to me, sounds the same as Alexis. It could be just a coincidence, but one interview with Alexis shows her as very outgoing cheerful and not camera shy, while in the other video two days later shows quite the opposite as she is frowning and appears to want to speak but appears fearful and unsure who to trust. When you look at the parents who constantly are touching her to make her aware of their presence makes you wonder what else is going on, or they're hiding?

If we look at Sandy Hook as a crime scene you have to look at everyone as suspects, it only makes sense as we do not want to let any evidence slip by. If we follow Nancy Lanza as a victim we see red flags popping up all over the place. Who is she and what part did she play in the school shooting?

She supposedly is described by some as a teacher, then substitute teacher, and teacher's aide. She is a teacher of special need students that go to the Sandy Hook school.

Nancy Lanza was also married to a man named Peter. There was a teacher that was asked if she knew Nancy Lanza, her name was Abbey Clements, she said she didn't know the last name. I think that statement by itself seems odd, as most teachers run in tight groups at all schools nationwide. What makes sense is that maybe the teacher might have known her under another name such as Annie Haddad. When you look at both Annie and Nancy they looked identical. Maybe because, they were the same person?

What is interesting about Annie is she too taught special needs there and had a student and looked like Nancy and was married to a man whose name too was Peter. They lived less than a mile apart from each other and traveled the same routes. Both women attended monthly mom night outs. Yet no one had ever seen them together at these meetings, or at the school at the same time?

If you look at the interview of the Soto family the children all were smiling with no sign of anger, sadness, or confusion, their sister was killed and yet they talk about her with total control and smiles as one giggles about the snow and sleet being their sister. Later you see on television a show about gun control and all of a sudden it's hard to speak and she points out there is no place in civilian life for an AR- assault rifle. That the weapon left baseball size holes in her sisters clothes that she saw. She states "personally we haven't seen any of the crime scene photos." How is she able to have seen her sisters clothes? Where was she at, when she saw them and would not the area have been locked down and secured?

She states the youngest boy Noah was shot eleven times, as they focus on large capacity guns. She was more emotional talking about guns than she was her sister.

Let us look at Mr. Robbie Parker, Emilie's father who is caught on the news camera actually before he was ready, as he is smiling and laughing having a good time. Here his six year old daughter is killed gun down violently, and he is smiling? In the video he tries to create tears and put on a persona of great emotional atrocity as he stands in front of the camera, yet he wishes to offer his condolences to not just the other victims families, but the shooters family as well? This is one day after the killing of his daughter. He stated the night before he spoke at the church special meeting? The same day his child was killed? He lived in Sandy Hook eight months and has since vanished?

If we look at Gene Rosen who became one of the focal points during the interviews. He was praised for saving and comforting some of the children, but the more we got to know him the more people doubted his entire story. Listening to Gene on one interview he said. "I thought I heard some gunshots, sometimes I hear deer hunters there shots were boom, boom with a pause between them, but these shots were like rapid shots that went very quick, boom, boom, boom."

The news reported that he took in many of the sandy hook children, which we find to then be only four out of seven hundred that went to that school? He took them in giving them stuffed animals and juice only then did he find out what happened and what they had been through according to FOX News.

You need to keep in mind that Gene Rosen is experienced in acting and is the CEO of Newtown Cable Advisory Council. He has appeared in several stage productions such as the Fantastic's. His story constantly changes on how he found the children.

One of the interviews he states he found the kids as he came out of the loft where he fed the cats, another one he was on his way to breakfast, then on one interview he said he saw the children through his window and finally he said he was walking home from breakfast and saw the children on his front lawn. There is the man Gene refers to talking kind of harshly to the children. On another video clip there is a woman guarding them? In some interviews he omits the man who is speaking kind of harshly to the children. He states he heard him say it's going to be alright, it's going to be alright, another one he says he hears him say it's going to be ok, it's going to be ok? How do you say those words harshly?

He stated there was a school bus driver with the kids and he invited them into his house. Would not the school bus driver inform him of what was going on? He stated he saw the six children and thought they were practicing a play or cub scouts? Were all these kids boys? Why would they be in his yard in the first place especially if they were practicing a play or cub scouts? Gene stated there were four boys and two girls, when the bus driver stated there were only four children.

He talks about what the children said, "We can't go back to the school, we can't go back to the school cause we don't have a teacher." He states, "I did not say anything. They just kept telling their story."

"Our teacher is dead. How can we go back we don't have a teacher." He said they kept talking about guns saying he had a big gun and a small gun. Then they talked about blood and said her name, referring to Ms. Soto the 27 year old teacher.

Here is another question, was Gene Rosen married? Does he have kids or grandchildren? If no why would he have stuffed animals in his house? There is a picture showing stuffed animals as though it was a staged photo. Why would you offer juice and stuffed animals in the first place? Gene was a former psychologist who had worked at the Fairfield Hills hospital, if we remember treated the mentally ill and criminally insane.

The school bus driver had the presence of mind to call his supervisor who must have had a contact list, so says Gene Rosen. Yet the school bus driver never had the presence of mind to bring it to Gene's attention upon being invited in to his house, instead he has to wait for the kids to tell him what happened fifteen, twenty minutes later? To me this is all so strange.

Gene stated he took the kids to the fire house. On one occasion he stated all the parents picked up the kids. The official report supposedly states that three children were picked up at Gene Rosen's house while the fourth kid was taken by both to the fire house. Gene mentions after the kids were picked up meeting with a woman by the name Scarlet Lewis, who is the mother of Jessie, who is another hero. Jessie was not just a victim but a hero as Jessie told the others to run while Adam was reloading his magazine.

The whole story is then debunked as video appears showing Gene walking around at the school at 10:30 am outside in the parking lot. He even did an interview at that time supposedly.

Lexxtext 526 Sandy Hook Video on YouTube, points out some interesting facts or assumptions, let's look at them and see where it may lead. Victoria Soto's Face book page was created four days prior to the shooting, on December 10,2012. Looking at the page you can see the date clearly. After this is discovered it is pulled down and fixed, repaired, or altered to match the event. The school nurse reported that Nancy Lanza was a nice kindergartner teacher, only to find out Nancy Lanza was never a school teacher there. There is the tribute video that was posted a month before the event on *VIMEO* web page. This video points out on the web page that viewers had traveled to that page back in November, way ahead of the shooting.

Let's look at Ryan Lanza and his identity confusion. Adam Lanza is found with his brothers identification card in the school. His brother Ryan is arrested and charged with the shooting before they find out it was not Ryan, but Adam. They report no one had seen Adam for three years and his brother hadn't seen him for two years. So how is it he has his brothers identification card?

There is FEMA L-366 planning for the needs of children in disasters, dated December 14, 2012 which is located on the Department of Emergency service & Public service, Emergency and management Homeland Security page, under what appears to be a scheduled calendar event. Interesting how this page was created so quickly, though if it were scheduled would make sense.

Emilie Parker was pictured with the President after the shooting. How could she be seen with the President after the shooting if she were one of the victims? Though this could explain why Robbie Parker was seen earlier smiling.

The SSDI (Social Security Death Index) states Adam Lanza died on December 12, 2012, two days before the shooting occurred. He killed twenty six and wounded one, making him very accurate with the death to injury ratio, being 26 to 1.

Apparently Connecticut has an assault rifle ban making it a class D felony. Where and how did Nancy Lanza acquire this weapon and how was Adam so skilled at using this weapon? Lt. Vance stated they had investigators looking at who every weapon belong to.

If Adam Lanza worked alone who was the guy that people saw in the woods handcuffed? A gentleman being interviewed stated a man walked by them saying he didn't do it. The gentleman being interviewed at the time stated, "he is in the back of the police car now. He was dressed in cameo pants and a dark jacket." Who was this guy and where is he now? According to the video you hear an officer say there are two shooters seen running. One was supposedly an off duty police officer from out of town, who just happened to get caught in the woods with only one road that lead in to the school? Was he the second shooter, or was he the one in the police vehicle?

Dawn Hochsprung had implemented a new security system prior to the shooting. Where is that video evidence at?

During all the videos you see not one victim being assisted by EMS, Law Enforcement. There is no blood, there is no video showing the children escaping.

What we do see is a staging area, one at the school and one at the church. We see no first responders moving in any manner as to help with victims, but rather just standing around as though they too are waiting for something to occur, rather than something that had occurred. We do not see bodies being brought out and loaded in ambulances. We see people walking around what at first looks just like confusion, but if you watch the many characters out there you will see they are walking routes that make this all appear to be more organized or scripted.

If you look at the 911 transcripts you will see the first caller called in at 9:35 am while another lady called in at 9:56 am. She was asking for ambulances and states she is outside and there appears to be none out there. Why would it take twenty minutes to get ambulances there? The dispatcher confirmed that they had been called because he stated yea they're coming. Then at the end of that he stated "We will have them come right now, ok." Why would you have to say that if they were already coming?

The second call came in approximately 17 seconds later by a caller who is distressed asking also for an ambulance. Third caller contacts 911 dispatch who answers, "Are you calling about the shooting in Newtown?" The caller states yes he is calling in about the shooting and states he is not injured. He states he is calling from the other school?

He states he was not involved, but that he just arrived and needs an ambulances on scene.
A caller told 911 dispatch to bring ambulance at base of school they're bringing victims out there. The ambulances never make it to the base of the school for some reason. A caller states we need assistance in the kindergartner room #3. This call was at approximately 10 am. Finally Ambulances arrive, but they still need more ambulances. They talk about how there are three injured yet at the time Adam Lanza was supposedly found dead as he took his own life. They do not mention anything about deceased personnel Which they would have had to see if they went by room 8 and room 10 according to the reporter. On another call they stated that they would need multiple ambulances and they had multiple KIA (killed in action).

When we look at United Way, we see they created a page to collect funds to help with the Sandy Hook travesty. The problem is they too jumped the gun and published their page well in advance of the Sandy Hook shooting. On the Google page there was a link that said (Sandy Hook School Support Fund) The document date page showed December 11,2012, Days before the shooting

The United Way of Western Connecticut said it rejects conspiracies theories, claiming it knew about December 14th Newtown Connecticut school shooting three days in advance and leveraged that knowledge to raise money. The New Way fund was actually established by the Newtown savings bank. The president and CEO of the bank is none other than John Trentacosta. The very same Trentacosta that owned the home next to the Lanzas.

There is aerial photos that show both the Lanza and the Trentacosta residences, which show tire tracks only in the driveway and grass of the Trentacosta property. The Lanza property, which was the area considered to be the area of interest due to Adam Lanza being the identified shooter, had no sign of activity to very little activity. There is also photos showing there was a road block at the end of the Trentacosta driveway. Why were they blocking that driveway and not the Lanza's driveway?

According to the reporters, John Trentacosta stated the fund was created due to countless requests. If we look at this as true, then when did he receive these requests? The reason I mention this is because this page was up and running at minimum on the same day of the shooting. That means someone developed the fund concept. You would have needed a planning meeting. You would also need to establish a fund account and coordinate with the United Way foundation. You would need to develop a campaign. You would need to create the art work web context and layout. Then publish this all on line.

This by itself makes it look as though there was knowledge of the incident going to take place ahead of when it did. You could almost call it a version of insider trading, whether it was for the purpose of gaining money, power, or political favors is hard to say, but not out of the realm.

If we look at the DNA evidence, we see more doubt and questions that once again don't add up.

There is the 22 Savage rifle, a Christmas card, an envelope, labeled for the students of Sandy Hook elementary school, the adhesive side of the stamp, and the exterior and interior of the door handles of the car located at the crime scene. They then got a DNA sample from Nancy Lanza's blood and from the person said to be the alleged Adam Lanza, getting a sample from his liver, so as to compare the DNA.

Based on the conclusion by the Forensic Science Examiner, Nancy was eliminated as a contributor and was eliminated from all items. Adam was eliminated as a contributor from the following items as well, however the only items that were swabbed that he could be connected to would be mixers. A DNA mixer is a sample that contains more than one individual's DNA.

You may ask how can more than one persons DNA be present if Adam Lanza acted alone? Was the DNA contaminated? On January 17, 2013 a hit was made to a convicted offender's DNA profile from the New York State Police Investigation Center. Who was this guy and how was he connected to the school or the Lanzas?

Eric Hurita the Forensic Examiner received an email asking questions of concern on the following questions. Would your conclusions indicate any of the following?

#1 Adam Lanza was not the shooter?

#2 Adam Lanza did not work alone?

#3 There was somebody else definitely involved?

#4 Was there any follow up on the convict from New York and if your findings put him as one of your suspects?

#5 Do you know the context of what was found inside the letter and why it was crucial to the investigation?

#6 The DNA swab that was taken from Adam Lanza was taken from his liver, is this common procedure?

#7 Were you ever brought into contact with the shooters body or was the liver swab the only source of Adam Lanzas DNA? According to the report he has yet to reply to the questions that we Americans want and deserve. Is the reason he can't answer because it is still ongoing, or is he afraid of self incrimination?

Let us look at Barbara Sibley, who during her interview stated her son had forgot something and when she arrived she could tell something wasn't right. She states there were eight or ten kids running towards the fire house. Who were these eight or ten kids? You will see Barbara arrives before the police, while a subject is still shooting. Were these kids running to the fire house or were these the ones that went to Gene Rosen's house?

She mentioned seeing a black hatchback and it was by the entrance and all the doors were open and jackets laying on the ground outside the doors of the vehicle. She saw another mom by the door and they mentioned how quiet it sounded then they noticed the broken glass and heard shots being fired.

She talks about seeing the kids leave as they are being escorted and she describes the event as organized as the children are being led by their teachers They show a picture with a girl running with her and in the interview she mentions her son. She states she sees her son but doesn't want him to see her hysterical? Really I would not care about anything but holding my child and blessing God for keeping him safe.

In the photo you see two women running with a girl and an officer, while another officer is seen in the background looking away from the building and what appears to be a third officer outside the door as well. Should not those three officers attack and eliminate the threat, rather than assist people that are temporarily safe in the parking lot?

The subjects vehicle is found to belong to a Christopher Rodia. So did Christopher report his car stolen, or was he one of the shooters as well?

One of the kids stated he heard a loud bang and thought something fell, then he heard another and thought it was gunshots. The teacher Katelyn Roy stated it happened during their meeting. When asked what happened she stated, the shooting. The first class room, she states and that she knew it was gunshots because it came from one of those guns that shoots over and over again. She stated she shut her door but didn't lock it? She stated she had them all get in the bathroom all 15 students plus herself. She stacked them in the small bathroom One of the boys stated he knew karate. She told him he needed to just remain where they were and remain calm.

One of the other kids stated during this ordeal that their teacher was reading them books and keeping them calm. Maybe this kid had a different teacher. The reporter on the video also mentioned there were candles that the children drew but when you looked at the candles they seemed to be taped on the windows so as to view them from outside, rather than from inside like you would expect. There are also shelves and file cabinets that are placed in front of them in the classroom. It appears odd to me and again appears staged for those that would be reporting to see this easily from outside, rather than from inside.

President Obama talked on his frustration with two and a half years left, stating "That this society has not been willing to take some basic steps to keep guns out of the hands of people who can do unbelievable damage. We are the only developed country on earth where this happens. It now happens once a week. It's a one day story, there is no place else like this." He talks about 26 year olds getting gunned down and this town couldn't do anything about it. Maybe he meant 26 people, referring to Sandy Hook, Connecticut.

This is why many people feel there is a hidden agenda and a secret that has been silenced from the rest of the public. The question then becomes, did anyone truly die? Was it a drill, or a drill gone bad? Were some people in on it, or was the whole town of Sandy Hook aware of this so called travesty?

The world deserves answers and answers come from questions that are asked or evidence that is uncovered, or revealed.

It appears the whole world is a stage

Are these the children that were savagely killed, or are they pictures of children that are now adult actors or accomplices with a cause?

Top picture An officer at an active shooting with his gun still in his holster and another officer with his back toward the school where the threat is supposedly coming from?

This is a statement by a Newtown resident. I thought it sounded very interesting.

"I live in Newtown CT. I will post no proof that I actually do live here. I am sure some have clicked away immediately at that confession and I do not blame them. You may or may not believe in what I am about to say and that is your choice.

I live in Newtown, I live about a mile from Sandy Hook Elementary on Riverside Road. I didn't go to work on December 14th, I called in sick, but I was playing hooky. I work at a restaurant in Danbury. I've spent the past two hours debating with myself if I should post this.

I could have posted this somewhere else, but this site seems to always pop up when the Sandy Hook shooting is concerned. I figured this was the most popular spot.

At around 8:50- 9:00am I had just finished eating my breakfast. I had Captain Crunch cereal. I had used the last of the milk. I decided to go get some at Caraluzzi's, because a friend of mine works there and I wanted to see what he was up to that evening. I get in my car at about 9:05am. I drive West on Riverside Road.

Before I even got to the fire house, I notice a lot of vehicles parked there. Way, way, way more than usual. A few were parked on both sides of the road. I figured they were just doing some kind of training or meeting.

However one thing did stick out and that was the fact that a lot of the vehicles were solid silver and solid black. I drove past the Fire House and the senior center is right next door less than a hundred yards. That parking lot was full as well, with some of the same vehicles mixed with trucks and what not. Vehicles you would expect to see in a parking lot. Still I thought nothing of it. I go get my milk.

I'm in Carraluzzi's and I've been talking to my friend for a right good while and we hear two police vehicles go by. About five minutes later we hear a Helicopter. Dave comes from the back and said, "They must be doing another drill again. Scanner said mass casualty incident involving children." Right then I figured out why all those vehicles were at the Fire House and Senior Center, because it is not the first time they have done that. They did a similar thing about a year ago, around this time of year and they are always doing car accidents that concern Hazmat, or whatever. I told them what I saw. Dave said, "Those cars sound like FBI cars. I guess they would be in on it.

So I figured I'd go, so I could catch a peek at what was going on. I get to town and couldn't go down Riverside Road. I told the officer that I live down there. He said, "No through traffic even for residents. Fire and Rescue and Police need to finish what they are doing."

I said, "What are they doing?" He said, "I would tell you if I knew and I have no clue.

The way it sounds they are running a drill, but we have been blocked from main traffic. I was just told to do traffic control and make sure no one goes down there."

Then I drive West to Danbury. I don't know why I just did. I was thinking to myself that is what I get for playing hooky from work. I had the radio on, but not loud enough to hear. I turn it up thinking maybe I could catch what is going on.

The local station mentioned nothing. At about 10:00am they break in from Sammy Hagar's "I can't drive 55". The guy said he was hearing reports of an incident at Sandy Hook Elementary and that surrounding schools were on lockdown , however "I called my wife because she works at Newtown Middle School and she said they weren't on lockdown and hadn't heard anything about an incident. So it appears information has been mixed up. We need to get our stuff together before they tell me to report".

I get to Danbury and I go into Global gas station to get a Mt. Dew. They had the TV on and I froze. They were watching coverage of the shooting. I just watched for thirty minutes frozen. Then I started thinking about all those cars at the Fire Department. I didn't know what to think about that. My brain just didn't work after watching the television.

Fast forward to right this second. I've read the theories and the stories and honestly, those are the only ones that add up. I live here and I do not know what the **** went on. This town is fairly small. I know right many people. I know Eugene Rosen. He is an extremely creepy bastard. I know Scarlett Lewis. But I would like to know where in the hell she got all that money to spruce herself up for interviews.

But the phelps and Parker families that seemed to be shoved down our throats, I have no idea in hell who they are. I've asked families and friends and they don't know who they are. I know a Robert Parker, but that sure as hell isn't him in the interviews. I just do not know what is going on.

The narrator says, what you are about to see are pages from a DHS (Department of Homeland Security)school shooting exercise plan, for a drill in Iowa that happened back in 2011. Looking at the context you see it states you need to arrive at your appropriate site fifteen minutes early. Wear the appropriate uniform and identification items. Sign in when you arrive.

Parking information and directions to each venue area are available from the Emergency Management Agency. Refreshments and portable water will be provided for all exercise participants throughout the exercise. Restroom facilities will be available at each venue.

Identification badges will be issued to the exercise staff. All exercise personnel and observers will be identified by agency uniforms and, or identification badges distributed by the exercise staff. They show a chart that describes these identification items. The badge colors represent the type of group as follows.

White: Exercise Director

Black: Exercise staff

Green: Controllers

Red: Evaluators

Orange: Actors

Yellow: Support Staff

Blue: Observers

Pink: Media Personnel

None: Players uninformed

Gray: Players in civilian Clothes

They showed pictures of stacks of bottled water set up inside one of the buildings in Newtown, Ct. The similarities are hard not to see. People gathered around and appear not as worried and confused, some personnel, possibly actors who are waiting to play their roles. We must remember shadows may depict more than one thing, creating an illusion of alternatives.

Let's look at Noah Pozner. A video on you tube shows him as a victim of a school shooting, not just of Sandy Hook Elementary, but as a victim of a school shooting in Pakistan. A British Boxer known as Amir Kahn visited a school in Pakistan where the Taliban massacred a hundred and fifty two people December 30,2014. He wanted to help scarred children return to school. I wonder if after being killed twice in two different countries, in two different school shootings Noah Pozner will be better prepared when the third school shooting takes place?

A picture is worth a thousand words, but a video speaks volumes. Noah's mother talked of her son and showed no sign of sadness. She tried to express herself as being numb to the News Media, as she portrays herself as deprived of the power of sensation. The problem is that Noah's mom wants us to share in her grieving process, while she has yet to show any signs of grieving. She is in front of the media on her own and was not forced to talk to the world, yet only speaks of her son, not about him. She too brings up guns rather than dealing with the shooters mental capacity.

She focuses on the AR assault rifle, describing it as though it were the cause of a tormented soul. When asked what her objective was with this, she stated "I haven't wrapped my head around it yet, I'm certainly no Statesman women." We will get back to that statement.

When Veronique Pozner talked on another interview she mentioned her son was a twin, but failed to mention that on the other interview. She was asked if she were afraid the death of her child might be all for nothing. She stated, "If we haven't reached critical mass. If there isn't some type of change in the national consciousness, by six and seven year olds being gun down in the sanctity of their own school then I think we have lost true North in this country. Our compass is broken and our priorities are completely off kilter." Even the interviewer was amazed at how she could talk about this without losing composer.

She states she's not always composed, that she cries every day, but she is just a private person. You could say in front of the camera she is always composed. If you remember Veronique stated , I'm certainly no Statesman women. You will see in one of the video links she Veronique Haller Pozner is a legal counselor for the Government of Switzerland. You can reach her directly in the Switzerland Embassy in Washington, DC.

Myself watching from the outside, have to ask are these things all anomalies? I am a Correctional officer I am trained in observation, what doesn't make sense today may mean everything tomorrow. Let me give a simple example using a picture of Emily with President Obama.

If the picture was taken after the shooting then how is she not reported as a survivor rather than a victim?

If the picture was taken before the shooting, you then have to ask why was the President there in the first place and again have to ask, what are the odds a child that gets to meet the President of the United States is then massacred.

Remember I stated we had to look at everyone as suspects, looking at Nancy Lanza's neighbors all of them seem to know nothing of her or of Adam. It is as though they were placed in a setting and went into hiding. No one seems to have ever been inside their home, no one?

Adam Lanza seemed to have disappeared in 2009 after his sophomore year. He is seen as being autistic yet is able to go to college having not gone through his junior and senior years in high school? He is able to wipe his computer so clean leaving nothing for the trained FBI investigators to find for evidence? He was able to kill 26 people leaving only one casualty, or three at the most? Maybe Adam Lanza is the real Glimmer Man. One girl named Israel who claims she is a classmate of Adams then states she never was actually in any of his classes? States she never seen him with anyone and that he would always sit alone at lunch. She mentioned Nancy Lanza as a kindergartner teacher when she wasn't.

The investigators mention that during the shooting spree that approximately 150 rounds were shot. Where are all the bullet holes, not including the ricochets which each shot can produce multiple shot markings.

Where is the video of Adam Lanza entering the school? They supposedly put up cameras and people had to be identified to get in to the school. It is interesting that the cameras we want to view we can't, while the other cameras are all over the place as everyone is strolling and patrolling to be seen and heard.

There is more videos out there in regards to this travesty. There are more questions that still need answers. During the one interview where Gene Rosen is talking you see a sign in the back flashing. The sign says, (Everyone must check in.) If we remember just a little while ago I had talked about that Iowa drill, where that was part of the protocol. You had to sign in when you arrived. When you watch videos of the Sandy Hook shooting you see people with ID's on. Find out more on those Id's and you might be amazed.

Charlotte Bacon, 2/22/06, female
- **Daniel Barden**, 9/25/05, male
- **Rachel Davino**, 7/17/83, female.
- **Olivia Engel**, 7/18/06, female
- **Josephine Gay**, 12/11/05, female
- **Ana M. Marquez-Greene**, 04/04/06, female
- **Dylan Hockley**, 3/8/06, male
- **Dawn Hochsprung**, 06/28/65, female
- **Madeleine F. Hsu**, 7/10/06, female
- **Catherine V. Hubbard**, 6/08/06, female
- **Chase Kowalski**, 10/31/05, male
- **Jesse Lewis**, 6/30/06, male
- **James Mattioli** , 3/22/06, male
- **Grace McDonnell**, 12/04/05, female
- **Anne Marie Murphy**, 07/25/60, female
- **Emilie Parker**, 5/12/06, female
- **Jack Pinto**, 5/06/06, male
- **Noah Pozner**, 11/20/06, male
- **Caroline Previdi**, 9/07/06, female
- **Jessica Rekos**, 5/10/06, female
- **Avielle Richman**, 10/17/06, female
- **Lauren Rousseau**, 6/1982, female (full date of birth not specified)
- **Mary Sherlach**, 2/11/56, female
- **Victoria Soto**, 11/04/85, female
- **Benjamin Wheeler**, 9/12/06, male
- **Allison N. Wyatt**, 7/03/06, female

Highlighted names appear in photo to the left from a photo taken February 2013 supposedly.

If these are the kids that were said to be killed we should be happy knowing they are all alive, however this is not the case, Americans are mad if this is what happened. We are tired of being lied to and deceived through deception. It is important to know we can trust our government, to know they will tell us the truth, the whole truth so help them God.

What is interesting is that the Sandy Hook School was tore down and destroyed. The home of Nancy and Adam Lanza was tore down and destroyed. The Virginia Tech School is still up and running after the shooting that took place on April 16,2007, killing 32 and wounding 17 others in two separate attacks. The Columbine School is still up and running after the April 20, 1999 shooting, killing 13 people. The McDonald massacre that took place in 1984, where forty bystanders were shot twenty one died, it too continued to operate.

When I compared some of the videos that are on You Tube you saw law enforcement moving with a purpose. You see students outside running, confused in chaos. Nothing on those other events show any signs of masking. Was the house and school destroyed to rid them of any evidence that might be left? Look at who was Commander in Chief at the time of these events? President Reagan 1984 McDonald Massacre, President Clinton 1999Columbine shooting, President George W. Bush 2007 Virginia Tech Massacre, President Obama 2012 Sandy Hook.

Francine Lopez Wheeler was identified as one of the parents of Benjamin Wheeler, who died during a mass shooting at Sandy Hook Elementary. She is a former personal assistant to the Finance Chair Woman and the Democratic Committee, Maureen White. Both Maureen and her husband Steven Rattner are members of the Council of Foreign Relations. Steven Rattner is known to have ruffled feathers with gun control advocate Michael Bloomberg according to the New York Times. Francine is an actor and singer. She became the first person other than Obama and Biden to deliver the White House weekly address. She identified herself as just a citizen.

Mark Barden, parent of alleged victim Daniel Barden is a lifelong entertainer, composer and musician, is shown on the White House. Gov. website. Barden leads policy on outreach efforts for sandy Hook promise, an organization that is committed to affecting policy in mental health, gun access, and enhanced security.

Nicole Hockley, mother of Dylan Hockley graduated from Trinity College, where she majored in English and theater. She publicly admitted regret of not continuing her acting career.

Jimmy Green father of Anna Marquez, is an entertainer and musician, both Jim and his wife are strong advocates for gun control. It appears that many of the children come from families who were in to the entertainment society, and or political parties with certain agendas such as gun control.

The truth is hard to find when you talk about Sandy Hook. Because it is the inconsistencies that make us want to take another look at this travesty. If this were a movie people would be asking for their money back as the script would be seen as poorly written. There are too many holes that leave doubt and suspicion.

The characters too appear to be lost in how they portray themselves. No one is seen with tears when they should. Their behaviors too appear fake and insincere. The timing is off in many areas that we had discussed. The children evacuating the school is not caught on tape or seen in photos other than two photos that appear to be staged using a couple of the same children.

The School is destroyed along with Nancy Lanza's home leaving again less evidence as to make it appear suspicious, getting rid of evidence that could later be looked at. I am not saying anyone is guilty or innocent, I am leaving that to you to decide. There is a lot more out there that could be looked at and if you are interested I have included some links that you can check out yourself and see what I saw.

Whatever our world has in store for us, we will be included.

The following are links to sites on YouTube that will get you thinking.

https://www.youtube.com/watch?v=X3aYQEJXJfo

https://www.youtube.com/watch?v=aQz1BIM9mLw

https://www.youtube.com/watch?v=oD0z275nQnM

https://www.youtube.com/watch?v=lV20DtXBuQs

https://www.youtube.com/watch?v=Hvhs5PWQW-o

https://www.youtube.com/watch?v=iCGDFUWVyG8

https://www.youtube.com/watch?v=3uZqtWWqD0E

https://www.youtube.com/watch?v=eYwPN5-FDoc

https://www.youtube.com/watch?v=ivt1KzZ9Bs8

https://www.youtube.com/watch?v=WmJO3ljBy7Y

https://www.youtube.com/watch?v=AUSJ6rqEWUY

Compare these videos to the videos of the other mass shootings and you will see the obvious. Let the truth speak for itself. If you are ready we will move on to our next story.

CHAPTER EIGHT

I covered the previous topic in my earlier book Deception This topic was brought up in this book because history has a way of bringing out the truth, as we have covered many areas in this book, from technology that seems unimaginable to a historic crime scene that has been more than just unfathomable, but deceivable by the news media.

Another story that I had written on previously was the Boston Marathon Bombing. If we look at that again along with the Sandy Hook Shooting then all that was, is and will be reported must be viewed carefully.

Remember as I told of the blanket, or afghan, each story is a stitch when carefully woven together create a fabric we might see as unimaginable. Is the fabric that holds our world together what we thought?

BOSTON BOMBING

On April 15, 2013 people had gathered both as spectators, as well as participants in an event known as the Boston Marathon. It was one of the prestigious of all marathons. People from all over the world came to take part in this event, but on this day it was not those that made it to the finish line that were going to be remembered. Instead those that died or were injured as innocent bystanders, along with those that would later be accused of perpetuating this whole ordeal.

At 2:49 pm EDT an explosion occurs near the finish line, twelve seconds later 210 yards further away a second explosion goes off. As two puffs of white smoke clouds and confuses those around the area.

Once again a beautiful day has been tarnished by those we call terrorists, those who are willing to kill others for a cause whether or not innocent lives are taken or not.

War has hit the streets of Boston, Massachusetts

America's news commentators are quick to report on the scene, as Americans listen for clarity of what happened.

This event was thought to be terrorist organized, catching us all off guard. Now years later this story too makes it in the pages of Deception, with questions of conspiracy, and fraud. Follow us as we look at this event closer.

The Federal Bureau of Investigation took over the investigation and three days later on April 18, 2013 they released the photos of Dzhokhar Tsarnaev and Tamerlan Tsarnaev publically. They are Chechen brothers. Shortly after the release of the photos, the subjects kill an MIT police officer, and hijack a civilian SUV. They ensue in a gun battle between themselves and the police in Watertown, Massachusetts.

Tamerlan Tsarnaev was shot several times in the gun battle, along with being rundown by the same vehicle they used to escape and evade with by his own brother. Tamerlan Tsarnaev was pronounced dead at the scene.

Thousands of law enforcement personnel swarmed the area of Watertown, covering a twenty block radius. Businesses, local institutions, along with public transportation were all shut down temporarily on a very large scale. People were asked to stay inside and to lock their doors.

Eventually Dzhokhar Tsarnaev was located behind a house in a boat unsure if armed or not he was shot and arrested then treated at a hospital.

CNN reports on the Boston Marathon bombing as they mention both Chechen brothers, Dzhokhar Tsarnaev and Tamerlan Tsarnaev were preparing across the Charles river with their own agenda. They had been viewing the magazine called Inspire which was created by Al-Qaida the terror group in the Arabian Peninsula. They look at a recipe on making a bomb in your mother's kitchen.

They talk about using pressure cookers and everyday items referring to nails for shrapnel. They were reported to have gone to the Home Depot in Cambridge to buy a soldering gun, fireworks that they used the gunpowder out of the fireworks, along with BBs to be placed in their homemade bomb.

As CNN covers the story they point out the location of the two brothers. You can see on the one video that the younger brother, Dzhokhar puts a phone to his ear as seconds later an explosion goes off as everyone looks towards the blast and he turns and heads the other way and again seconds later the next bomb goes off right in the same area that the younger brother was just standing. You can see in the video that Dzhokhar leaves the area quickly and then you see him twenty four minutes later in a store buying milk as though nothing has happened. He clearly knows what has happened, as he is seen on the tape looking towards the first explosion.

In his defense though other people were in there shopping as well. The difference is that if he initiated the explosions he appears to be cold and heartless. In the video you can also see clearly that he is without his backpack that he had twenty four minutes earlier. Dzhokhar's friends identified him as normal. Dzhokhar arrived in 2002 speaking very little English at the age of eight years old. He grew up and became a wrestling captain on his Cambridge wrestling team. He was liked by many.

His brother Tamerlan Tsarnaev, was seven years older than him and was having a harder time adjusting to American culture.

Unlike Dzhokhar who was well liked and had many friends Tamerlan had none. Their sisters too were failing with marriages and their parents with earning a living. Tamerlan found a sense of belonging in radical Islam. Him and his mother received a visit from the FBI who interviewed them on concerns of radicalism, but nothing came of it due to lack of evidence.

Tamerlan heads to Dagestan next to Chechnya. Dagestan is torn by ethnic violence and extremism. It is during this visit that something happens that connects with Tamerlan and he passes information and idealism back to his brother, Dzhokhar. When Tamerlan returned six months later he is noticed as more radical and him and his brother listen to radical Islamist messages.

As the forensic teams sift through the crime scene area looking for evidence, they find parts of the pressure cookers, parts of the back pack and a twisted fuse. They also go over and over the videos looking at things that stands out. One of those items was none other than Dzhokhar who is seen on the video during the bombing not flinching and moving away in the opposite direction very clearly. They point out that Dzhokhar text a friend telling him if he wants he can take what's there, referring to his property.

The brothers appear to have no plan thought out, making you wonder on that note by itself if they had concocted this on their own or if they had taken direction from someone else. On the video you see where the brothers, or the subjects approach the officers vehicle and then are seen running from the vehicle in the same direction that they came from.

Tamerlan and his brother hijack an SUV telling the driver they were the ones that did the Boston Marathon Bombing, and that they shot a police officer.

The driver escapes from his own vehicle when they pulled into get gas. The driver runs into another convenience store and asks them to call 911, stating that he was hijacked by the guys that did the Boston bombing. Police respond and use the GPS to track the vehicle to Watertown where the shootout took place.

The shootout was intense as Tamerlan runs out of ammunition and WAs tackled by officers. His brother Dzhokhar sees this and heads the vehicle towards the officers that were wrestling with his brothers as the police are being warned to get out of the way. Dzhokhar continues driving towards them, but misses the officers and runs his brother over instead killing him.

Dzhokhar escapes temporarily and the town is put on lockdown. He is eventually found in a boat in one of the residents back yard. He was shot and arrested, then treated medically. In the boat Dzhokhar left a note that he had written reported by CNN it said, " The U.S. Government is killing our innocent civilians… I can't stand to see such evil go unpunished."

He wrote as well the following saying, "We Muslims are one body, you hurt one you hurt us all." He also stated, "I'm jealous of my brother… I do not mourn because his soul is very much alive." Dzhokhar was charged with thirty federal counts, including conspiring to use a weapon of mass destruction. Dzhokhar pleads not guilty. In the end he is found guilty of all the charges.

In another video you can see someone had videoed the bombing on their supposed cell phone. On the video when they slow it down you do not see any blood. What you do see is a woman that is wearing red and during the crisis is not bloody in two separate pictures shown later you see her in a wheel chair with an injured left hand and face, to me looks very suspicious.

In the one video you can see a hand on a leg as he is directing the person in the wheel chair by jerking the leg in the intended direction. If this were real then the person who is handling the leg, I would say is no professional. They point out the reason for shaking the leg was to make sure it was securely attached for the wheel chair ride in front of the cameras.

There is the controversy with Jeff Bauman's left leg being amputated long before the bombing as they show two pictures of his left leg with a bulge, along with another picture that shows his leg with no bruising or irritation, but instead what looks like a healthy healed leg. On the very same video they point out the tattoos on his arms saying they could have been erased. That seems like a lot of thinking for this event but all should be looked at. They want you to focus on the length of the stubs, which are longer in some pictures and shorter in others.

They point out on the video his left thigh is too long, relative to his arm. They point out his left leg if they add a foot to it, which they do using photo ops on the video. They say if his left leg is too long his right leg is even longer. They point at his right knee cap protruding out in a certain position.

They do bring up the question why does he look so calm and not in shock, which can be argued either way. But he does look well poised and not in any pain like I would think one would be in after losing a leg, or two.

There is also the Marathon hero who wore a cowboy hat, his name was Carlos Arredondo. At the beginning of the video you see Carlos with another gentleman holding up a sign that says, (Latinos remaking America). They then mention that Carlos had come from Costa Rica across Arizona in 1980 and obtained his green card by testifying against the people he paid to bring him here. They mention his ex-wife is Melida Arredondo, who is a peace activist. While it is true America is made up of diversity it does seem like activists, as well as lobbyists are uniquely woven together whether by accident or on purpose in these chaotic crisis.

Carlos was identified as a hero who had jumped and cleared fences along with helping rescue and save Jeff Bauman, who had lost both limbs as they mention he had his arteries severed in his legs. They point out if both arteries are cut then it increases the severity of the injury to include death by bleeding out. On the video you would expect to find Carlos attending to Jeff who needs immediate attention, but we do not see that.

What we see is a man that tries to appear to be helping while he clinches hold of the American flag. It makes me wonder if he is holding the flag so we know what team he is on. What would make a person focus so much on the flag when there are those that need our help? It is possible he sees himself helping a cause rather than a person.

In the video you do see him handed something that he sticks in his pocket from someone in a yellow jacket. He does take the time to look at it as he puts it in his pocket which is later identified as a bloody flag. Carlos mentions as he is interviewed that he dropped the flag, which we do not see on video at any time.
Though again if he did drop the flag why not continue on helping Americans in dire straits?

Carlos states that he picked up Jeff and put him in the wheelchair. This too would seem odd when there are people all over that could help assist him, and would if this was real. Jeff too stated that Carlos was the one that picked him up off the ground. What they forget to mention is the other people that were caught on camera lifting Jeff up into a wheelchair. It is possible Jeff was in shock at that moment. Carlos had tunnel vision, focusing only on what he thought was his mission to get Jeff in the wheelchair. There were two other people besides himself that took him by wheelchair out of the crisis area, or crime scene, which ever you prefer to call it.

He mentions he is part of Red Cross and is somewhat trained in disasters, yet the word tourniquet he had to have said to him by his ex-wife. I know we all have forgotten words before, so yes it is possible. I am trying to be fair to both sides of the conspiracies.

On the video he says he is a member of the Red Cross as he holds up a badge. Could this be one of those badges we heard about in the Sandy Hook story? In the video you see him bend down and attach what appears to be a new badge to his chain as he struggles getting it over his cowboy hat. He is seen with two different badges, why?

If you listened to enough of the witnesses that were there they will tell you they saw FBI agents and bomb sniffing dogs early on. They will tell you the FBI were conducting a bomb exercise there at the same time. Is this just a coincidence or was it part of the exercise?

They show Carlos phone, where he showed a picture of Jeff laying in agony. Why would you be focused on taking pictures if not for a cause, or some sadistic purpose of pleasure, or monetary gains?

On the Boston Hero or NOW FRAUD video you see areas where people are not standing or sitting, one minute, then you see them laying and positioned in a manner that looks staged.

Carlos mentioned that Jeff's shirt was on fire when it is clear it is not, was not, or will ever be, yet he sticks with this statement. There is also the changing of the cowboy hats where they show you two distinctly different hats one that says Costa Rica on it in the front while the other does not.

You may think does a hat really matter? Well it does if it is missing evidence such as blood. They show a picture of his hat during the April bombing and one month later. I guess it took them a month to alter the hat.

I guess the hat matters enough to put it on eBay and sell it for, a buy it now price of One Million Dollars. Just another perfect example of raising money from what is suppose to be a real crisis. I wonder how much he wants for the bloody flag?

His shirt is not without critiquing, as we see no blood on the sleeve till later in one interview, yet we have seen his sleeve in clear view as he is assisting Jeff in the wheelchair. You would think too, if he lifted Jeff as he said he did there would be more blood on his shirt. The truth is that he had help and the video substantiates that.

His pants too showed no blood on them during the interview and later in the video from Amy Goodman's he has blood all over them. It again makes me think of Sandy Hook as the hero becomes the Ponzi. Where so much attention is put on them it is hard to keep the facts straight and the scenes in order.

You see during the explosion the flags were just fine, but listening to Carlos Arredondo the flags were all gone, destroyed. The truth is the flags appeared to have not even been torn, shredded, punctured by any shrapnel. What are the odds of that being possible considering all the flags that were there in the bombs path?

Let's look at Carlos's history. On August 25, 2004 Carlos was outside his home in Hollywood, Florida when three marines arrived to give him the news his son Alexander had been killed in Iraq. He loses it and torches the van the marines arrived in setting himself on fire by mistake as well.

Two days after the van attack Carlos wakes up to two strangers by his bedside. He called them officers and stated this case was taken over by the FBI and Homeland Security. He admits he thought he was in trouble for targeting U.S. Government property. He mentioned everything was taken care of.

Carlos stated after his son died President Bush signed a security order, giving citizenship to parents of the sons and daughters that died in the Iraq War. He stated, "I received the first one granted to a parent, with the help of Senator Kennedy." Remember he is an illegal immigrant. He is not deported or sent to Guantanamo, he was not charged with destroying government property, or given a bill for his hospital stay. He did however receive a first class trip from Florida to Massachusetts for three, for his son's funeral. This is supposedly when all the press coverage began.

The media focuses on Carlos and Melida, and not his first wife Victoria who actually raised his children. They portray Melida as the grieving mother. It is possible that the media was just duped by a woman who tries to slide in and replace Victory who never seems to get mentioned. We know Melida as an activist, but the video points out opportunist as well. When Melida says they were responsible for their son's service, this is a perfect example of deception. You then have to ask why does all this matter unless you have an objective, or agenda. Who is it they are wanting to impress or influence? Maybe they are using all this to pay a debt back to someone with elite authority?

He who looks away from the obvious is just as blind
On a radio interview Carlos and Melida couldn't even agree on whether or not they had talked with the family of the guy they saved?

We are only humane but in my entire life I have never known anyone to allow a news media, or photographer in to take pictures of those grieving over their loved ones.

I have to sound the alarm of skepticism on this issue. A respected Time Magazine that is political doing a story on this? Was it the loss of a Marine, or focused on an illegal who got a second chance?

Carlos has been paid with free travel, free publicity and notoriety he has been honored all over at games and has spoken at political conventions does he not standout as someone very special?

On this same video they point out how racism is used to separate this country, because a country divided can't come together to protect itself, which is what the elite want supposedly. To prove that statement they show a clip from none other than the famous Opera Winfrey who stated the following, "There are still generations of people, older people who are born and bred and marinated in that prejudice of racism and they just have to die."

Why do these people have to die? Should we not educate others on racism and know if we can educate the old the young will follow? I know this all seems to come out of nowhere, but it is the idea that the bombing at the Boston Marathon was fake as we have covered some interesting points that would be hard to dispute, along with the links that will show you with your own eyes.

On another video you see a picture of the victim Jeff Bauman, who is identified as Nick Vogt. I could be wrong on the last name, but he is a former U.S. Army officer who lost his legs in Kandahar, Afghanistan. He was with the 1st Stryker Brigade 25th infantry division in 2011. This is what they report on the video.

Let's look at this next picture and tell me what you see.

Another victim that has died twice, you would think with today's technology they would do makeup to at least mask the obvious. I am not sure if the crisis actors are getting better or worse, but Americans are waking up to the trickery. If this is true who can and will explain this? I would love to hear any reasonable explanation.

On this video they claim that former CIA agent, Robert David Steele has declared that the Sandy Hook and the Boston Marathon bombing were false flags, where a staged event is presented to influence a targeted audience.

Nicole Brannock Gross reportedly broke both her tibia and fibula on both her legs, while she sits calmly for twenty minutes waiting to be treated or showcased.

There is one CNN crisis actor that has been caught four times as GMN reports, she was caught at the actual Boston marathon bombing, then in Watertown, and also in Sandy Hook shooting. This was on a video called Crisis actor caught again 4th time. The last one was the NTSB train derailment.

You would think that they could find new actors, instead of using the very same ones. Some of this information came from the Pete Santilli Show. On another video they show once again a different girl that they say was at the Sandy Hook, Boston Marathon, and Aurora. They state the reason for crisis actors is to throw out a perception to say things without giving details. That tells me it is not about perception, but deception.

In one video Victoria McGrath is seen being carried by a first responder She was supposedly injured when the bomb went off just a few feet from her. Her shoe doesn't look like it has even been involved in the same crisis. This was on the video titled Boston Marathon bombing survivor killed. She was supposedly killed in a car accident overseas. Not sure if we will find out later she never died, but was doing a crisis over seas.

There is Bruce Mendelsohn who stated he was one or two blocks away from the bombing when it knocked him out of his seat. So did it physically blow him out of his seat or did he get startled and jump out voluntarily?

Americans want to know the truth, just like when news commentators report the news and stage their backgrounds to appear somewhere else.

In this one video clip two reporters are talking back and forth as though they are both in completely different locations. However the vehicles in the background show a different scenario where at best they are a hundred feet from each other. True fakery at its worst. This was done and is on the link I have included in this book. All you need to do is check out these links and make your own educated decision.

There was Charles Jacobs, news reporter who too faked his location and reported on a story where he never was, this was posted on August 27, 2009 and was part of CNN. I had to laugh at the antics they used while reporting as one person puts on a helmet and the other one puts on a gas mask. It is obvious they never thought this through if there is a chemical attack the guy with the helmet is in trouble. They also point out the blue screen which is used like a green screen in editing, letting you be placed anywhere in the world while in one room.

In my Changing America books I had brought up Hillary Clinton being a survivor of sniper fire that never happened, that she said did, till she was caught in a lie and then wanted to blame it on sleep deprivation.

Maybe that is what the news media has when they report things that aren't real. The news media should be held accountable for all it reports, and to correct itself as soon as they catch their own mistakes or others bring it to their attention. This is where accountability becomes an important tool with policing our Truth Sayers, or those portraying the truth as respected news media, politicians and other officials that Americans look up to.

Remember at the beginning of the marathon FBI agents were spotted on top of buildings and had bomb sniffing dogs present as they were telling the people they had nothing to worry about that this was a training exercise.

There are a lot of red flags that are waving at us in this horrific event, but nothing points to just one person but a mass of different people at different levels throughout our country. We do need to be vigilant, when it comes to the next crisis. We need to look at each event as who has the motive, means, and opportunity to do these things. We need to look at those who will be the victims and those who will play the part of the predator and follow their history to see if anything stands out.

Since we have identified a few heroes for you, it seems only fair we go into the next story covering a hero we all know BATMAN, as we get ready to look at the Batman Shooting that happened in the theater in Aurora.

When only truth is spoken, lies cannot be heard.

CHAPTER NINE

We are nearing the end of this book. We have covered technology that seems beyond our belief, from beaming ourselves in two places through cameras to actually traveling to another dimension, or universe. We have our military that is continually getting more advanced with the aid of robots, lasers, and force fields. We have Doctors, Professors, and other high ranking personnel exposing our world to HARP and CERN, where earthquakes and tsunami are manmade, along with portals that our government admits to. There is the Mandela Effect where what was is now no more, as words and letters have been changed or completely erased. When words from our bible are deleted then our Christianity is in stake. We have again mind control that too was admitted to by our government using MK ULTRA.

We have biometrics that are being used in everyday life from Disney World to airports. We see that cloning is not just in animals but in plants and soon if not already in us. GPS is all around us as well, cars, phones, internet, to the stock exchange. In fifty years our phones have emerged immensely.

There is the debate of our world being flat verses round. There is the Hallow earth debate as well with again a very distinguished naval officer who told of this. I brought back two chapters of my last book *Deception* because our history is important to our future, but what if all we remember is or has already been changed? Science has proven we have the capability to do extraordinary tasks with amazing results.

Let's not forget that the sun was brought up not to brighten our day but to expose the conspiracy it is not hot as they point out that the closer you get to the sun the hotter it should get yet the temperature decreases the farther out in space you get

While all we have talked about in this book are impressive, scary, confusing, topics, we have to know there is so much more out there that is hidden from us. When they can make cities appear above in the clouds above another city for all to see we have to wonder of the things they have yet to show us.

Are we at the end of days? Are we already doomed? Only God knows what is in store for us. I believe he wants us to believe in him and to love one another. He wants us to treat others the way we would want to be treated.

I know there are those we are angry at, but even those people he wants us to forgive. For if you cannot forgive them how can he forgive you? We must remember to ask ourselves, who are we to question God?

We are here on this world because of God. We only have to follow our footsteps backwards to see this. We talked about aliens, even Bigfoot which many see as a hoax. If God created us could he not have created Bigfoot? What about the devil himself? For if we believe in God you then must believe in Satan. We will all be judged one day, this I believe. Now you may believe differently and I can respect that.

The part that is scary is knowing that we are not helping any when we go and change things that only God should be able to do. I know this is argumentative in the idea we need to continue to improve and stay ahead of the other countries, but that is what they are doing as well. Someday we will reach the top and there will be nowhere to go.

CHAPTER TEN

What will be written in the next chapter of our lives?

Check out the book that started all of the Changing America series. Beginning with the racial riots of yesterday. With Michael Brown and the town called Ferguson in Missouri.

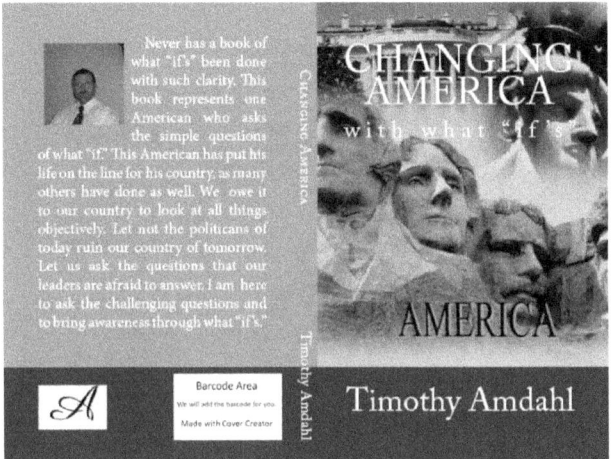

Changing America Volume II continues on and brings to light the outcome of some of the earlier topics and a continuation of what I think can only be seen as an idiotic movement through political correctness at its worse.

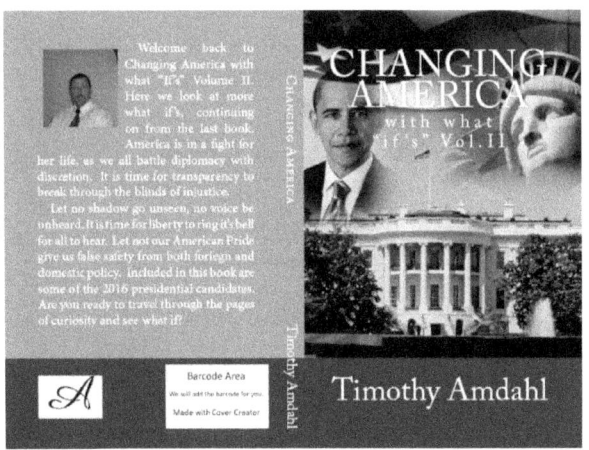

The final book Changing America 2016 is the beginning of the end of the last eight years of total chaos It covers every president and looks at the presidential candidates for 2017. I knew America would pick the right one as America was tired of all the giving our country has done, to the denial of Muslim extremists.

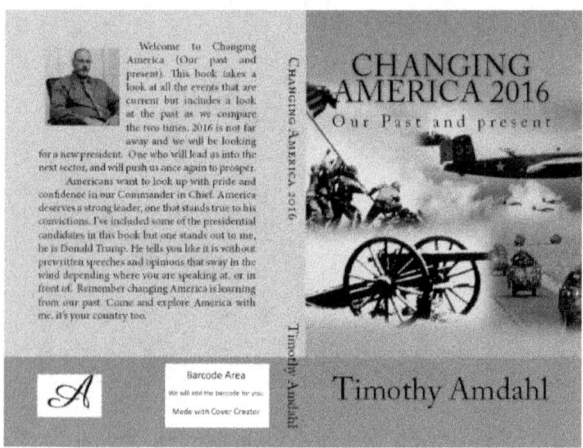

I hope you have enjoyed this book and will look at these other books as well. Our country needs to be observant along with the rest of the world. God gave us life he gave us a home here on this planet called earth. If nothing else we must never lose our faith, for God is here, has been and will continue to be when we are all gone.

Thank you Lord and thank all of you who took the time to read this.

God Bless America and our World

ABOUT THE AUTHOR

Timothy J. Amdahl grew up in a small town, called Estherville, Iowa. He graduated in 1981 and served two years with the Army and then transferred in to the United States Marine Corps. He was honorably discharged in 1987 after serving four and a half years in the Marine Corps. He has worked as a youth counselor for four years helping the children of our future.

He is married with four children and is currently working for the Illinois Department of Corrections as a Correctional officer, having already served sixteen years. He is a proud American who only wishes to unite our country once again in these troubling times.

The Ever Changing World is focused on bringing out our awareness as we look to our future. Hopefully through knowledge and awareness we will be better ready for tomorrow. I know the question becomes, who really controls tomorrow, as we focus on what we are morally and ethically to do and what God wants us to do. Let us all pray we are doing the right thing.